ANNOTATED
DICTIONARY
OF
CONSTRUCTION
SAFETY
AND HEALTH

ANNOTATED DICTIONARY

OF

CONSTRUCTION

SAFETY

AND HEALTH

Charles D. Reese

From the Handbook of OSHA Construction Safety and Health
by Charles D. Reese and James V. Edison

LEWIS PUBLISHERS

Boca Raton London New York Washington, D.C.

Library of Congress Cataloging-in-Publication Data

Reese, Charles D.
    Annotated dictionary of construction safety and health / Charles D. Reese, James V. Eidson.
      p.   cm.
    ISBN 1-56670-514-2 (alk. paper)
    1. Building—safety measures—Dictionaries. I. Eidson, James V. II. Title.
TH443.R432 1999
690′.22′03—dc21   99-042159
                                                      CIP

The material in this book was taken from the Handbook of OSHA Construction Safety and Health by Charles D. Reese and James V. Eidson.

No claim to original U.S. Government works
International Standard Book Number 1-56670-514-2
Library of Congress Card Number 99-042159
Printed in the United States of America   2 3 4 5 6 7 8 9 0
Printed on acid-free paper

PREFACE

The construction industry has always been viewed as a unique industry. Although there are many aspects which are the same as in other industries, it certainly has its share of unique hazards. The intent of this dictionary is to provide a ready tool which can be used to address the occupational safety and health issues faced by those working in the construction industry; this includes contractors, workers, safety and health professionals, project managers, suppliers, and manufactures of equipment and materials.

As an easy-to-use guide for safety and health in the construction industry, this dictionary becomes the foundation upon which to build stronger safety and health initiatives within the construction industry, while intervening and preventing jobsite deaths, injuries, and illnesses.

Charles D. Reese, Ph.D.

ABOUT THE AUTHOR

CHARLES D. REESE

For over twenty years Dr. Charles D. Reese has been involved with occupational safety and health as an educator, manager, or consultant. In Dr. Reese's early beginnings in occupational safety and health, he held the position of industrial hygienist at the National Mine Health and Safety Academy. He later assumed the responsibility of manager for the nation's occupational trauma research initiative at the National Institute for Occupational Safety and Health's (NIOSH) Division of Safety Research. Dr. Reese has had an integral part in trying to assure that workplace safety and health is provided for all those within the workplace. As the managing director for the Laborers' Health and Safety Fund of North America, his responsibilities were aimed at protecting the 650,000 members of the laborers' union in the United States and Canada.

He has developed many occupational safety and health training programs which run the gamut from radioactive waste remediation to confined space entry. Dr. Reese has written numerous articles, pamphlets, and books on related safety and health issues.

At present Dr. Reese is a member of the graduate and undergraduate faculty at the University of Connecticut, where he teaches courses on OSHA regulations, safety and health management, accident prevention techniques, industrial hygiene, and ergonomics. As Associate Professor of occupational safety and health, he coordinates the bulk of the safety and health efforts at the university and Labor Education Center. He is often called upon to consult with industry on safety and health issues and also asked for expert consultation in legal cases.

TABLE OF CONTENTS

Introduction

Safety and the potential risk of serious injuries in the construction industry are often interrelated or even synergistic with other hazards associated with construction work. As efficiently as one might try to address each safety hazard on a construction worksite or within the construction industry, some hazards or some facet of a construction hazard may be overlooked. Often the unique work experiences which a contractor, health and safety professional, supervisor, or worker has had make them aware of a hazard or risk which could not be foreseen without the special knowledge of the jobsite or work experience, which those individuals within the construction industry possess. It is often impossible to duplicate that practical experience on the written page.

This dictionary covers the most common types of construction risks or hazards which may impact the safety and health of construction workers. It addresses each topic in alphabetical order but does not delve into subtle nuances which can occur on a construction worksite or project. It also attempts to address the OSHA requirements for protecting workers on construction jobsite and projects.

One of the complaints often voiced by many is that practical real life construction is often lacking in discussions on construction safety and its hazards. This dictionary meshes the regulations, common sense, and practical construction work aspects as much as is logically possible within these few pages.

The safety hazards are presented in alphabetical order. This is to facilitate quick reference to the specific topic of interest to the reader. Each narrative relevant to a specific construction safety risk or hazard should not be the only source that one consults regarding that hazard. Use the appropriate regulations, OSHA-generated materials, industry-generated materials, and manufacturers' materials to try to assure a complete understanding of the hazards, the risks, and potential intervention strategies to mitigate accidents and injuries.

ABRASIVE GRINDING (1926.303)

All grinding machines are to be supplied with sufficient power to maintain the spindle speed at safe levels under all conditions of normal operation. Grinding machines are equipped with safety guards in conformance with the requirements of the American National Standards Institute, B7.1-1970, Safety Code for the Use, Care, and Protection of Abrasive Wheels. The safety guard covers the spindle end, nut, and flange projections. The safety guard is mounted so as to maintain proper alignment with the wheel, and the strength of the fastenings exceeds the strength of the guard.

Floor-stand and bench-mounted abrasive wheels used for external grinding are provided with safety guards (protection hoods). The maximum angular exposure of the grinding wheel periphery and sides are not more than 90 degrees except when work requires contact with the wheel below the horizontal plane of the spindle, then the angular exposure shall not exceed 125 degrees. In either case, the exposure begins not more than 65 degrees above the horizontal plane of the spindle. Safety guards are to be strong enough to withstand the effect of a bursting wheel. Floor and bench-mounted grinders are provided with work rests which are rigidly supported and readily adjustable. Such work rests are kept at a distance not to exceed
one-eighth inch from the surface of the wheel. (See Figure 1 for an example of abrasive grinder.)

All abrasive wheels should be closely inspected and ring-tested before mounting to ensure that they are free from cracks or defects. All employees using abrasive wheels are protected by eye protection equipment in accordance with the requirements except when adequate eye protection is afforded by eye shields which are permanently attached to the bench or floor stand.

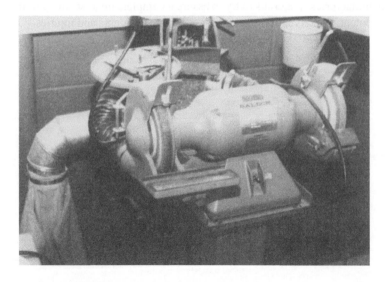

Figure 1. Abrasive grinder

AERIAL LIFTS (1926.556)

Aerial lifts are required to meet design and construction guidelines of Vehicle Mounted Elevated and Rotating Platforms (ANSI A 92.2-1969). Aerial lifts include the following:

1. Extensible boom platforms.

2. Aerial ladders.

3. Articulating boom platforms.

4. Vertical tower.

5. A combination of any of the previous.

They may be constructed of a wide variety of materials and may be powered or manually operated. These aerial lifts can be field modified if approved by the manufacturer, testing laboratory, or certified engineer. (See Figure 2 for an example of an aerial lift.)

Prior to movement of an aerial lift, the boom or ladder is to be secured and locked into place to assure that the outriggers are properly stowed. Prior to use, lift controls are to be tested each day. Aerial lifts used to transport workers must have both lower and upper controls which are of easy access to operators. The lower controls should be designed to override the upper controls. All controls are to be labeled according to their function. The lower controls should never be used unless permission is given by the operator in the basket, or for an emergency. Only authorized workers shall operate a lift. The manufacturer's load limits for the boom or basket should be followed. When outriggers are set on pads or solid surfaces, the brakes are to be set, as well as wheel chocks in place. Aerial lift trucks are not to be moved when workers are in the basket unless equipment is so designed for such movement.

Figure 2. Aerial lift with worker wearing a safety belt

Never alter the insulating capacity of the boom or an aerial lift. All electrical and hydraulic bursting factors must comply with ANSI A92.2-1969. All components must have a bursting factor of 2 to 1.

Workers in a basket should have a body harness with a lanyard attached to the boom or basket but never attached to an adjacent pole or other equipment. While working in a basket, the worker is to remain completely within the basket. Climbers should never be worn in an aerial lift.

AIR RECEIVERS (1926.306)

Compressed air receivers and other equipment are used for providing and utilizing compressed air when performing operations such as cleaning, drilling, hoisting, and chipping. All new air receivers, which are installed after the effective date of the air receiver regulations, are to be constructed in accordance with the 1968 edition of the A.S.M.E. Boiler and Pressure Vessel Code Section VIII, and all safety valves must also be constructed, installed, and maintained in accordance with Section VIII.

Air receivers should be installed such that all drains, handholes, and manholes therein are easily accessible. Under no circumstance is an air receiver buried underground or located in an inaccessible place. A drain pipe and valve are to be installed at the lowest point of every air receiver in order to provide for the removal of accumulated oil and water. Adequate automatic traps may be installed in addition to drain valves. The drain valve on the air receiver is to be opened and the receiver completely drained, frequently, and at such intervals as to prevent the accumulation of excessive amounts of liquid in the receiver.

Every air receiver must be equipped with an indicating pressure gage (so located as to be readily visible) and with one or more spring-loaded safety valves. The total relieving capacity of such safety valves is such as to prevent pressure in the receiver from exceeding the maximum allowable working pressure by more than 10 percent. No valve of any type is placed between the air receiver and its safety valve or valves. Safety appliances such as safety valves, indicating devices, and controlling devices are to be constructed, located, and installed so that they cannot be readily rendered inoperative by any means, including the elements. All safety valves are tested frequently and at regular intervals to determine whether they are in good operating condition.

ALARMS (1926.159 AND .602)

Although an alarm is not normally considered a hazard, the absences of a unique alarm system can be a hazard. An alarm system needs to be present on the jobsite. There needs to be a unique alarm signal which alerts workers to specific hazards or actions which need to be taken when an emergency exists, such as evacuation. These alarms should be recognizable and different from commonly used signals such as for breaks, lunch, or quitting time.

All emergency employee alarms installed to meet a particular OSHA standard are to be maintained, tested, and inspected. This applies to all local fire alarm signaling systems used for alerting employees, regardless of the other functions of the system. The employee alarm system must provide a warning for necessary emergency action, as called for in the emergency action plan, or for reaction time for safe escape of employees from the workplace or the immediate work area, or both. The employee alarm is to be capable of being heard above ambient noise or light levels by all employees in the affected portions of the workplace. Tactile devices may be used to alert those employees who would not otherwise be able to recognize the audible or visual alarm. The employee alarm must be distinctive and recognizable as a signal to evacuate the work area or to perform actions designated under the emergency action plan.

The employer is to explain to each employee the preferred means of reporting emergencies, such as manual pull box alarms, public address systems, radios, or telephones. The employer shall post emergency telephone numbers near telephones, employee notice boards, and other conspicuous locations when telephones serve as a means of reporting emergencies. Where a communication system also serves as the employee alarm system, all emergency messages have priority over all nonemergency messages.

The employer is to establish procedures for sounding emergency alarms in the workplace. For those employers with ten or fewer employees in a particular workplace, direct voice communication is an acceptable procedure for sounding the alarm, provided all employees can hear the alarm. Such workplaces need not have a back-up system. The employer must assure that all devices, components, combinations of devices, or systems constructed and installed, are approved. Steam whistles, air horns, strobe lights or similar lighting devices, or tactile devices that meet the requirements of this section, are considered to meet this requirement for approval.

The employer must restore all employee alarm systems to normal operating conditions as promptly as possible after each test or alarm. Spare alarm devices, and components that are subject to wear or destruction, are to be available in sufficient quantities at each location for prompt restoration of the systems. The employer must make sure that all employee alarm systems are maintained in operating condition except when undergoing repairs or maintenance.

A test of the reliability and adequacy of nonsupervised employee alarm systems is made every two months. A different actuation device is used in each test of a multi-actuation device system so that no individual device is used for two consecutive tests. The employer is to maintain or replace power supplies as often as is necessary to assure a fully operational condition. Back-up means of alarm, such as employee runners or telephones, are provided when systems are out of service.

The employer assures that all supervised employee alarm systems are tested at least annually for reliability and adequacy. The employer assures that the servicing, maintenance, and testing of employee alarms are done by persons trained in the designed operation and functions necessary for reliable and safe operation of the system. The employer makes sure that the manually operated actuation devices, used in conjunction with employee alarms, are unobstructed, conspicuous, and readily accessible.

ARC WELDING AND CUTTING (1626.351)

Burns and electricity are the hazards confronted by the arc welder. Thus, only manual electrode holders which are specifically designed for arc welding and cutting, and are of a capacity capable of safely handling the maximum rated current required by the electrodes, are to be used. Any current-carrying parts passing through the portion of the holder which the arc welder or cutter grips in his hand, and the outer surfaces of the jaws of the holder, must be fully insulated against the maximum voltage encountered to ground. All arc welding and cutting cables are to be completely insulated, flexible, capable of handling the maximum current requirements of the work in progress, and take into account the duty cycle under which the arc welder or cutter is working (see Figure 3).

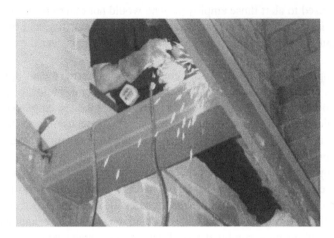

Figure 3. An arc welder

Cable that is free from repair or splices must be used for a minimum distance of ten feet from the cable end to which the electrode holder is connected, except cables having standard insulated connectors, or splices with insulating quality equal to that of the cable, are permitted. When it becomes necessary to connect or splice lengths of cable one to another, substantial insulated connectors of a capacity at least equivalent to that of the cable shall be used. If connections are effected by means of cable lugs, they must be securely fastened together to give good electrical contact, and the exposed metal parts of the lugs must be completely insulated.

Cables in need of repair are not used. When a cable becomes worn to the extent of exposing bare conductors, the exposed portion must be protected by means of rubber and friction tape or other equivalent insulation. A ground return cable is to be of safe current carrying capacity equal to or exceeding the specified maximum output capacity of the arc welding or cutting unit which it services. When a single ground return cable services more than one unit, its safe current-carrying capacity is to equal or exceed the total specified maximum output capacities of all the units which it services.

Pipelines containing gases or flammable liquids, or conduits containing electrical circuits, are not to be used as a ground return. For welding on natural gas pipelines, the technical portions of regulations issued by the Department of Transportation, Office of Pipeline Safety, 49 CFR Part 192, Minimum Federal Safety Standards for Gas Pipelines, apply.

When a structure or pipeline is employed as a ground return circuit, it must be determined that the required electrical contact exists at all joints. The generation of an arc, sparks, or heat at any point causes rejection of the structures as a ground circuit. When a structure or pipeline is continuously employed as a ground return circuit, all joints are to be bonded, and periodic inspections must be conducted to ensure that no condition of electrolysis or fire hazard exists by virtue of such use.

The frames of all arc welding and cutting machines need to be grounded either through a third wire in the cable containing the circuit conductor, or through a separate wire which is grounded at the source of the current (see Figure 4). Grounding circuits, other than by means of the structure, are checked to ensure that the circuit between the ground and the grounded power conductor has resistance low enough to permit sufficient current to flow to cause the fuse or circuit breaker to interrupt the current. All ground

Figure 4. Grounding for an arc welder

connections are to be inspected to ensure that they are mechanically strong and electrically adequate for the required current.

Workers are to be instructed in the safe means of arc welding and cutting as follows:

1. When electrode holders are to be left unattended, the electrodes must be removed and the holders are so placed or protected that they cannot make electrical contact with employees or conducting objects.

2. Hot electrode holders are not dipped into water; to do so may expose the arc welder or cutter to electric shock.

3. When the arc welder or cutter has occasion to leave his work or stop work for any appreciable length of time, or when the arc welding or cutting machine is to be moved, the power supply switch to the equipment is opened.

4. Any faulty or defective equipment is reported to the supervisor.

Whenever practical, all arc welding and cutting operations are shielded by non-combustible or flameproof screens which will protect welders and other persons working in the vicinity from the direct rays of the arc.

BARRICADES (1926.202)

These structures provide a substantial deterrent to the passage of individuals or vehicles. Barricades for the protection of workers must conform to the National Standards Institute D6.1-1971, *Manual on Uniform Control Devices for Streets and Highways*, Barricades Section. The most familiar type of barricade is the infamous "jersey barrier" which is made of reinforced concrete, but barricades can be made of other mate-

Figure 5. The use of jersey barriers as barricades

rials, if it provides an adequate protection factor. See Figure 5 for an example of a barricade.

BARRIERS

Barriers are used to warn workers and others of existing hazards and are usually not as substantial as barricades. They may be made of rope, wire ropes, warning tapes, or plastic fencing. A warning sign should accompany such barriers, either detailing the hazard or cautioning not to enter the area cordoned off by the barrier. See Figure 6 for an example of a barrier.

Figure 6. A barrier marked with caution tape

BATTERIES (1926.441)

Batteries of the unsealed type should be located in enclosures with outside vents or in well-ventilated rooms and arranged so as to prevent the escape of fumes, gases, or electrolyte spray into other areas. Ventilation is to provide diffusion of the gases from the battery and to prevent the accumulation of an explosive mixture, especially hydrogen gas. When racks and trays are used to store batteries, they should be substantial and treated to make them resistant to the electrolyte. An example of a poorly designed battery charging area can be seen in Figure 7. Floors should be of acid resistant construction unless protected from acid accumulations. Face shields, aprons, and rubber gloves should be provided for workers handling acids or batteries. Also, facilities must be provided for the quick drenching of eyes and the body and be within 25 feet (7.62 m) of battery-handling areas.

Facilities must be provided for flushing and neutralizing spilled electrolyte and for fire protection. Designated areas need to be provided for the purpose of battery charging installations and they must be located in areas designated for that purpose and charging apparatus must be protected from damage by trucks. When batteries are being charged, the vent caps must be kept in place to avoid electrolyte spray. Vent caps are to be maintained in functioning condition.

CHANGE ROOMS (1926.51)

When workers are required to wear PPE, there is the possibility of contamination with toxic materials. When this is the case, there must be a change room with storage for street clothes, so the contaminated PPE can be removed or disposed.

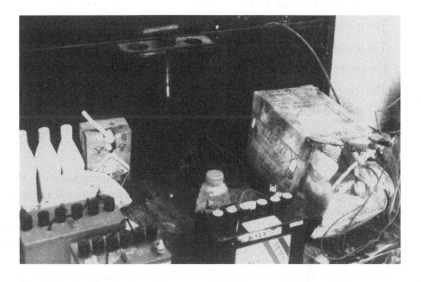

Figure 7. Poorly designed battery charging area

COMPRESSED AIR, USE OF   (1926.302)

Compressed air is not to be used for cleaning purposes, except when reduced to less than 30 psi, and then only with effective chip guarding and personal protective equipment which meets the requirements of Subpart I. The 30 psi requirement does not apply for concrete forms, mill scales, and similar cleaning purposes.

COMPRESSED AIR, WORKING UNDER

When working under compressed air, no employee is permitted to enter a compressed air environment until a physician has examined and reported him/her to be physically qualified to engage in such work. At least one physician is to be available at all times, while work is in progress, to provide medical supervision of employees engaged in compressed air work. If an employee is absent from work for 10 days or more, they must be reexamined by the physician prior to returning to work in a compressed air environment.

Employees continuously employed in compressed air will be reexamined by the physician within one year to determine if they are still physically qualified to engage in compressed air work. Examination records will be maintained by the physician.

A fully equipped first aid station will be provided at each tunnel project, regardless of the number of persons employed. A medical lock must be established and maintained in working order whenever air pressure in the chamber is increased above the normal atmosphere.

An identification badge will be provided to any employee working in a compressed air environment. The badge must give the employee's name, address of the medical lock, telephone number of the licensed physician for the compressed air project, and contain instructions for rushing wearer to the medical lock. The badge must be worn at all times – off the job, as well as on the job.

Effective and reliable means of communication (e.g., bells, whistles, or telephones) will be continuously maintained at the following locations:

- The working chamber face.
- The working chamber side of the man lock near the door.
- The interior of the man lock.
- Lock attendant's station.
- The compressor plant.
- The first aid station.
- The emergency lock (if required).
- The special decompression chamber (if required.)

A record must be kept of employees who work under compressed air. This record must be kept for each eight-hour shift and is to be located outside the lock and near the entrance. Every employee going under air pressure for the first time must receive instruction on how to avoid excessive discomfort. Except in an emergency, no employee working in compressed air will be permitted to pass from the working chamber to atmospheric pressure until after decompression.

Lock attendants who are in charge of a man lock will be under the direct supervision of the physician. They must be stationed at the lock controls on the free air side during the period of compression and decompression, and remain at the lock control whenever employees are in the working chamber or main lock. Lighting in compressed air chambers must be by electricity exclusively and use two independent sources of supply. The emergency source must automatically operate in the event of failure of the regular source. Lighting must not be less than 10 foot-candles on any walkway, ladder, stairway, or working level.

Firefighting equipment must be available and in good working condition at all times.

COMPRESSED GAS CYLINDERS (1926.350)

Compressed gas cylinders have the potential to become a guided missile if broken or damaged and will release a tremendous amount of energy. Thus, compressed gas cylinders should be treated with respect. Most compressed gas cylinders are approximately 1/4 inch in thickness, weigh 150 pounds or more, and are under some 2,200 pounds per square inch (psi) of pressure.

When transporting, moving, or storing compressed gas cylinders, valve protection caps should be in place and secured. When cylinders are hoisted, they must be secured on a cradle, slingboard, or pallet. (See Figure 8 for an example of a compressed gas lifting cradle.) They should never be hoisted or transported by means of magnets or choker slings. The valve protection caps are never used for lifting cylinders from one vertical position to another. Bars should not be used under valves or valve protection caps to pry cylinders loose when frozen. Warm, not boiling, water is used to thaw cylinders loose.

Compressed gas cylinders can be moved by tilting and rolling them on their bottom edges, but care must be taken to not drop or strike them together. This is especially true while

Figure 8. Cradle for lifting compressed gas cylinders

transporting them by powered vehicles since they must be secured in a vertical position. In most cases, regulators are removed, and valve protection caps put in place, before cylinders are moved. (See Figure 9 which shows cylinders that have been secured against falling.)

Oxygen cylinders that are put in storage are to be separated from fuel-gas cylinders or combustible materials (especially oil or grease). They must be separated by a minimum distance of 20 feet (6.1 m), or by a noncombustible barrier of at least 5 feet (1.5 m) high; this barrier must have a fire-resistance rating of at least one-half hour. Cylinders should be stored in assigned places that are away from elevators, stairs, or gangways; in areas where they will not be knocked over or damaged by passing or falling objects; or in areas where they are subject to tampering by unauthorized persons.

Cylinders are to be kept far enough away from the actual welding or cutting operation so that sparks, hot slag, or flame will not reach them. When this is impractical, fire resistant shields must be provided.

Also, cylinders should be placed where they cannot become part of an electrical circuit. Cylinders containing oxygen, acetylene, or other fuel gas must not be taken into confined spaces. Cylinders, whether full or empty, are not to be used as rollers or supports. No damaged or defective cylinders are to be used.

All employees should be trained in the safe use of fuel gas. The training should include how to crack a cylinder valve to clear dust or dirt prior to connecting a regulator. The worker cracking the valve should stand to the side and make sure the fuel gas is not close to an ignition source. The cylinder valve is always opened slowly to prevent damage to the regulator. For quick closing, valves on fuel gas cylinders are not opened more than 1 1/2 turns. When a special wrench is required, it is left in position on the stem of the valve. It is kept in this position while the cylinder is in use, so that the fuel gas flow can be quickly shut off in case of an emergency. In the case of manifolded or coupled cylinders, at least one such wrench is always available for immediate use. Nothing is placed on top of a fuel gas cylinder when it is in use; this may damage the safety device or interfere with the quick closing of the valve.

Fuel gas is not used from cylinders unless a suitable regulator is attached to the cylinder valve or manifold. Before a regulator is removed from a cylinder valve, the cylinder valve is always closed and the gas released from the regulator. If, when the valve on a fuel gas cylinder is opened, there is found to be a leak around the valve stem, the valve is closed and the gland nut tightened. If this action does not stop the leak, the use of the cylinder is discontinued, and it shall be properly tagged and removed from the work area. In the event that fuel gas should leak from the cylinder valve rather than from the valve stem, and the gas cannot be shut off, the cylinder is properly tagged and removed from the work area. If a regulator attached to a cylinder valve will effectively stop a leak through the valve seat, the cylinder need not be removed from the work area. But, if a leak should develop at a fuse plug or other safety device, the cylinder is removed from the work area.

COMPRESSED GAS WELDING

When using fuel gas and oxygen for compressed gas welding, certain precautions are necessary.

Fuel gas and oxygen manifolds should have the name of the substance they contain written in letters at least 1 inch high. This should be painted on the manifold or on a sign permanently attached to it. Fuel gas and oxygen manifolds are to be placed in safe, well-ventilated, and accessible locations. They are not to be located within enclosed spaces. The manifold hose connections, including both ends of the supply hose that leads to the manifold, are to be designed so that the hoses cannot be interchanged between fuel gas and oxygen manifolds and supply

Figure 9. Compressed gas welding set up with chain-secured gas cylinders

header connections. Adapters are not to be used to permit the interchange of hoses. Hose connections must be kept free of grease and oil. When not in use, manifold and header hose connections are to be capped. Nothing is to be placed on top of a manifold when in use, which will damage the manifold or interfere with the quick closing of the valves (see Figure 9).

Fuel gas and oxygen hoses are to be easily distinguishable from each other. The contrast may be made by different colors or by surface characteristics readily distinguishable by the sense of touch. Oxygen and fuel gas hoses are not to be interchangeable. A single hose having more than one gas passage is not to be used. When parallel sections of oxygen and fuel gas hoses are taped together, not more than 4 inches out of 12 inches may be covered by tape. All hoses in use, carrying acetylene, oxygen, natural, or manufactured fuel gas, or any gas or substance which may ignite or enter into combustion, or be in any way harmful to employees, should be subject to inspection at the beginning of each working shift. Defective hoses must be removed from service. Hoses which have been subject to flashback, or which show evidence of severe wear or damage, must be tested at twice the normal pressure to which it is subject, but in no case less than 300 psi Defective hoses, or hoses in doubtful condition, should not be used.

Any hose couplings must be of the type that cannot be unlocked or disconnected by means of a straight pull without rotary motion. Any boxes used for the storage of gas hoses must be ventilated. Hoses, cables, and other equipment are to be kept clear of passageways, ladders, and stairs.

Clogged torch tip openings should be cleaned with suitable cleaning wires, drills, or other devices designed for such purpose. Torches in use are to be inspected at the beginning of each working shift for leaking shutoff valves, hose couplings, and tip connections. Defective torches are to be removed from service. All torches are to be lighted by friction lighters or other approved devices, and not by matches or from hot work.

Oxygen and fuel gas pressure regulators, including their related gauges, must be in proper working order while in use. Oxygen cylinders and fittings are kept away from oil or grease. Cylinders, cylinder caps and valves, couplings, regulators, hoses, and apparatus are kept free from oil or greasy substances and are not handled with oily hands or gloves. Oxygen is not directed at oily surfaces, greasy clothes, or within a fuel oil or other storage tank or vessel. Oxygen must never be used for ventilation. Do not use oxygen to blow off clothes or clean welds; serious burns or death can occur

Additional rules – For additional details not covered in this subpart, applicable technical portions of American National Standards Institute, Z49.1-1967, Safety in Welding and Cutting, apply.

CONCRETE CONSTRUCTION (1926.701)

Compliance with concrete construction regulations will help prevent injuries and accidents that occur too frequently during concrete and masonry construction. No construction loads are placed on a concrete structure or portion of a concrete structure unless the employer determines, based on information received from a person who is qualified in structural design, that the structure or portion of the structure is capable of supporting the loads.

Employees are required to wear a safety belt or equivalent fall protection, when placing or tieing reinforcing steel more than six (6) feet above any working surface. All protruding reinforcing steel, onto which workers could fall, must be guarded to eliminate the hazard of impalement. (See Figure 10.)

During concrete construction, workers should not be permitted behind the jack during tensioning operations, and signs and barriers must be erected during tensioning operations to limit access to that area.

Concrete buckets, equipped with hydraulic or pneumatic gates, must have positive safety latches or similar safety devices installed to prevent premature or accidental dumping. Concrete buckets are to be designed to prevent concrete from hanging up on the top and sides. Workers are prohibited from riding in concrete buckets and from working under concrete buckets while the buckets are being elevated or lowered into position. If at all possible, elevated concrete buckets are to be routed so that no employee, or the fewest number of employees, are exposed to the hazards associated with falling concrete buckets. (See Figure 11.)

There are specific requirements for equipment used in concrete construction work. Bulk storage bins, containers, and silos must be equipped with conical or tapered bottoms and

Figure 10. Protected rebar to guard against impalement

Figure 11. Concrete bucket in use on construction site

a mechanical or pneumatic means of starting the flow of material. No employee should be permitted to enter storage facilities unless the ejection system has been shut down, locked out, and tagged to indicate that the ejection system is not to be operated. Lifelines/harness are be used when workers must entry bins, etc. that need to be unclogged. Workers are not permitted to perform maintenance or repair work on equipment (such as compressors, mixers, screens, or pumps used for concrete and masonry construction activities) where inadvertent operation of the equipment could occur and cause injury, unless all potentially hazardous energy sources have been locked out and tagged. Tags must read "Do Not Start," or similar language, to indicate that the equipment is not to be operated.

Concrete mixers, with one cubic yard or larger loading skips, must be equipped with a mechanical device to clear the skip of materials; guardrails must also be installed on each side of the skip.

Powered and rotating type concrete troweling machines, that are manually guided, must be equipped with a control switch that will automatically shut off the power whenever the hands of the operator are removed from the equipment handles. (See Figure 12.) Concrete buggy handles may not extend beyond the wheels on either side of the buggy.

Figure 12. Worker using a powered trowel

Figure 13. Construction worker using a bull float

Concrete pumping systems which use discharge pipes must have pipe supports designed for 100 percent overload. Compressed air hoses, used on concrete pumping systems, must be provided with positive fail-safe joint connectors to prevent separation of sections when pressurized. Sections of tremies and similar concrete conveyances are secured with wire rope (or equivalent materials), in addition to the regular couplings or connections. No worker is permitted to apply a cement, sand, or water mixture through a pneumatic hose unless the employee is wearing protective head and face equipment.

Bull float handles, used where they might contact energized electrical conductors, are constructed of nonconductive material or insulated with a nonconductive sheath, whose electrical and mechanical characteristics provide the equivalent protection of a handle constructed of nonconductive material (see Figure 13). Masonry saws are guarded with a semicircular enclosure over the blade and a method for retaining blade fragments is incorporated in the design of the saw's semicircular enclosure.

CONCRETE CAST-IN-PLACE (1926.703)

Formwork is designed, fabricated, erected, supported, braced, and maintained so that it will be capable of supporting, without failure, all vertical and lateral loads that are anticipated to be applied to the formwork (see Figure 14). Formwork must be designed, fabricated, erected, supported, braced, and maintained in conformance with strength specifications. These specifications must include the appropriate drawings or plans, including all revisions for the jack layout, formwork (including shoring equipment), working decks, and scaffolds that need to be available at the jobsite.

All shoring equipment (including equipment used in reshoring operations) is to be inspected prior to erection to determine that the equipment meets the requirements specified in the formwork drawings

Shoring equipment, which is found to be damaged and its strength reduced to less than the required standards, must be immediately reinforced. All erected shoring equipment should be inspected immediately prior to, during, and immediately after concrete placement; if found to be damaged or weakened, after erection, such that its strength is reduced to less than that required, it must immediately be reinforced.

The sills for shoring are sound, rigid, and capable of carrying the maximum intended load. All base plates, shore heads, extension devices, and adjustment screws are in firm contact, and secured when necessary, with the foundation and the form. Eccentric loads on shore heads and similar members are prohibited unless these members have been designed for such loading.

Whenever single post shores are used one on top of another (tiered), the employer must comply with the following specific requirements, in addition to the general requirements for formwork:

1.   The design of the shoring is prepared by a qualified designer and the erected shoring is inspected by an engineer qualified in structural design.

2.   The single post shores are vertically aligned.

3.   The single post shores are spliced to prevent misalignment.

4.   The single post shores are adequately braced in two mutually perpendicular directions at the splice level. Each tier is also diagonally braced in the same two directions.

Adjustment of single post shores to raise formwork is not made after the placement of concrete. Reshoring is erected, as the original forms and shores are removed, whenever the concrete is required to support loads in excess of its capacity. The steel rods or pipes on which jacks climb, or by which the forms are lifted, are to be specifically designed for that purpose and adequately braced when not encased in concrete.

All forms must be designed to prevent excessive distortion of the structure during the jacking operation. All vertical slip forms are provided with scaffolds or work platforms where employees are required to work or pass.

Jacks and vertical supports should be positioned in such a manner that the loads do not exceed the rated capacity of the jacks. Whenever failure of the power supply or lifting mechanism could occur, the jacks or other lifting devices are to be provided with mechanical dogs or other automatic holding devices to support the slip forms. The form structure is to be maintained within all design tolerances specified for plumbness during the jacking operation, and the predetermined safe rate of lift is not to be exceeded.

Any reinforcing steel for walls, piers, columns, and similar vertical structures must be adequately supported to prevent overturning and collapse. Also, employers are to take measures to prevent unrolled wire mesh from recoiling. Such measures may include, but are not limited to, securing each end of the roll, or turning over the roll.

Forms and shores (except those used for slabs on grade and slip forms) are never removed

Figure 14. Formwork for cast-in-place concrete

until the employer determines that the concrete has gained sufficient strength to support its weight and superimposed loads. Such determination shall be based on compliance with one of the following:

1.   The plans and specifications stipulate conditions for removal of forms and shores, and such conditions have been followed.

2.   The concrete has been properly tested with an appropriate ASTM standard test method designed to indicate the concrete compressive strength, and the test results indicate that the concrete has gained sufficient strength to support its weight and superimposed loads.

Reshoring is not removed until the concrete being supported has attained adequate strength to support its weight and all loads placed upon it.

Finally, formwork which has been designed, fabricated, erected, braced, supported, and maintained in accordance with Sections 6 and 7 of the *American National Standard for Construction and Demolition Operations Concrete and Masonry Work*, ANSI A10.9-1983, shall be deemed to be in compliance with the provision of 1926.703(a)(1).

CONFINED SPACES (1910.146 AND 1926.21)

Improper entry into confined spaces has resulted in approximately 200 lives lost each year. Confined spaces are not of adequate size and shape to allow a person to enter easily, have limited openings for workers to enter and exit, and are not designed for continuous human occupancy. Examples of confined spaces are storage tanks, silos, pipelines, manholes, and underground utility vaults. (See Figure 15 for an example of a confined space.) In evaluating confined space accidents in the past, certain scenarios predominated the event. These included the failure to recognize an area as a confined space; test, evaluate, and monitor for hazardous atmospheres; train workers regarding safe entry; and establish rescue procedures.

With the promulgation of the "Permit-Required Confined Space Entry" standard (29 CFR 1910.146) (Note: This standard is not directly applicable to construction.), all of these failures are addressed. First, a written and signed permit is required prior to entry, if the space contains or has the potential to contain a hazardous atmosphere (oxygen deficient, flammable, toxic); if it contains materials that could engulf the entrant; if the space's configuration could cause the

Figure 15. A typical confined space with a barrier and warning sign

Permit - Required Confined Space Decision Flow Chart

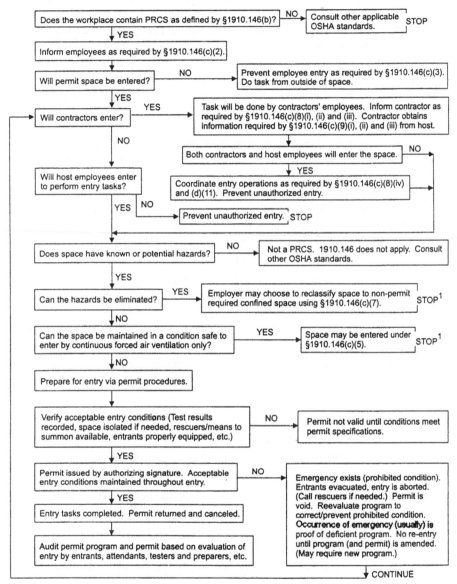

1 Spaces may have to be evacuated and re-evaluated if hazards arise during entry.

Figure 16. Determining if a permit-required space exists. Courtesy of OSHA

entrant to become entrapped or asphyxiated by converging walls; or if it contains any other recognized serious safety (electrical) or health (pathogen bacteria) hazard. This regulation requires that all permit-required spaces be identified, evaluated, and controlled.

The flow chart in Figure 16 can be used to determine whether a space is a permit-required confined space.

If a permit entry confined space exists then a written permit is needed which details and assures that procedures for entry exist; appropriate equipment and training for the authorized entrant(s) is provided; entry supervisors and attendants are trained and present; a written/signed entry permit (see Figure 17 for an example of a entry permit) exists prior to entry; trained and available rescue personnel exists (over one-half of the deaths in confined spaces are rescuers); the space has been posted with warning signs; barriers have been erected; and personal protective and rescue equipment are provided.

Work in confined spaces can be safely accomplished by paying attention to the key areas prior to entry. These are to identify, evaluate/test and monitor, train, and plan for rescue. Do not enter confines spaces without prior approval or a permit. Follow confined space entry procedures. Never enter a space unless you are trained. Wear required confined space entry PPE. Make sure, prior to entry, that arrangements have been made in case rescue is needed.

One of the least recognized, and most dangerous hazards on a construction jobsite, is working in confined spaces. Entry into confined spaces without the proper precautions could result in injury and/or impairment, or death due to

- An atmosphere that is flammable or explosive.

- Lack of oxygen to support life.

- Toxic materials, that upon contact or inhalation, could cause injury, illness, or death.

- General safety hazards such as steam, high pressure systems, or other work area hazards.

In an effort to prevent injury or death when working in confined spaces, the contractor should implement and enforce the following safe work procedures:

1. A confined space entry permit must be completed and signed by an authorized person prior to entry into a confined space.

2. A hazard evaluation must be conducted before any work is started in a confined space.

3. Technically competent personnel (i.e., industrial hygienist, safety specialist, etc.) must test the atmosphere within the confined space with an appropriate gas detector and approved oxygen testing equipment before employees enter.

4. If combustible gases are detected, employees are prohibited from entering the confined space until the source has been isolated and the space flushed or purged to less than 10% of the lower explosive limit.

5. If an oxygen-deficient atmosphere (less than 19.5% by volume) is present, positive ventilation techniques, including fans and blowers, may be used to increase the oxygen content. If further testing indicates the atmosphere is still oxygen deficient, Self-Contained Breathing Apparatus (SCBAs), or other air supplied respiratory protection will be provided.

```
┌─────────────────────────────────────────────────────────────────────┐
```

ENTRY PERMIT

PERMIT VALID FOR 8 HOURS ONLY. ALL COPIES OF PERMIT WILL REMAIN AT JOB SITE UNTIL JOB IS COMPLETED

DATE:_____ SITE LOCATION and DESCRIPTION _____

PURPOSE OF ENTRY _____

SUPERVISOR(S) in charge of crews   Type of Crew Phone #
_____
_____

COMMUNICATION PROCEDURES _____

RESCUE PROCEDURES (PHONE NUMBERS AT BOTTOM) _____

* BOLD DENOTES MINIMUM REQUIREMENTS TO BE COMPLETED AND REVIEWED PRIOR TO ENTRY*

REQUIREMENTS COMPLETED	DATE	TIME
Lock Out/De-energize/Try-out	___	___
Line(s) Broken-Capped-Blanked	___	___
Purge-Flush and Vent	___	___
Ventilation	___	___
Secure Area (Post and Flag)	___	___
Breathing Apparatus	___	___
Resuscitator - Inhalator	___	___
Standby Safety Personnel	___	___
Full Body Harness w/"D" ring	___	___
Emergency Escape Retrieval Equip	___	___
Lifelines	___	___
Fire Extinguishers	___	___
Lighting (Explosive Proof)	___	___
Protective Clothing	___	___
Respirator(s) (Air Purifying)	___	___
Burning and Welding Permit	___	___

Note: Items that do not apply enter N/A in the blank.

**RECORD CONTINUOUS MONITORING RESULTS EVERY 2 HOURS

CONTINUOUS MONITORING**	Permissible	
TEST(S) TO BE TAKEN	Entry Level	
PERCENT OF OXYGEN	19.5% to 23.5%	
LOWER FLAMMABLE LIMIT	Under 10%	__ __ __ __ __ __
CARBON MONOXIDE	+35 PPM	__ __ __ __ __ __
Aromatic Hydrocarbon	+ 1 PPM * 5PPM	__ __ __ __ __ __
Hydrogen Cyanide	(Skin) * 4PPM	__ __ __ __ __ __
Hydrogen Sulfide	+10 PPM *15PPM	__ __ __ __ __ __
Sulfur Dioxide	+ 2 PPM * 5PPM	__ __ __ __ __ __
Ammonia	*35PPM	__ __ __ __ __ __

* Short-term exposure limit:Employee can work in the area up to 15 minutes.

+ 8 hr. Time Weighted Avg.:Employee can work in area 8 hrs (longer with appropriate respiratory protection).

REMARKS:_____

GAS TESTER NAME & CHECK #	INSTRUMENT(S) USED	MODEL &/OR TYPE	SERIAL &/OR UNIT #
_____	_____	_____	_____
_____	_____	_____	_____

SAFETY STANDBY PERSON IS REQUIRED FOR ALL CONFINED SPACE WORK

SAFETY STANDBY PERSON(S)	CHECK #	CONFINED SPACE ENTRANT(S)	CHECK #	CONFINED SPACE ENTRANT(S)	CHECK #
_____	_____	_____	_____	_____	_____
_____	_____	_____	_____	_____	_____

SUPERVISOR AUTHORIZING – ALL CONDITIONS   SATISFIED_____

DEPARTMENT/PHONE _____

AMBULANCE _____ FIRE _____ Safety _____ Gas Coordinator _____

Figure 17. Sample written entry permit. Courtesy of OSHA

Figure 18. Confined space worker retrieval system with attendant

6. When toxic or chemical materials are detected or suspected, the following actions should be taken:
    - Any piping that carries or may carry hazardous materials to the confined space, will be isolated.

    - Empty the hazardous substance from the space until safe limits are reached.

    - Provide adequate ventilation and personal protective equipment for the eyes, face, and arms, if welding, burning, cutting, or heating operations, which may generate toxic fumes and gases, are performed.

    - Employees must wear eye and other appropriate protective equipment to prevent possible contact with corrosive materials.

7. An emergency plan of action that provides alternate life support systems and a means of escape from confined spaces must be developed and communicated to all employees engaged in work in confined spaces.

8. For evacuation purposes, each employee entering a confined space should wear a safety belt equipped with a lifeline, in case of an emergency. (See Figure 18 for example of a confined space retrieval system.)

9. Emergency equipment (e.g., lifelines, safety harnesses, fire extinguishers, breathing equipment, etc.) appropriate for the situation should be ready and immediately available.

10. All persons engaged in the confined space activity must receive training in the use of the life support system, rescue system, and emergency equipment.

11. An attendant, trained in first aid and respiration, must remain outside the entrance to the confined space. The attendant must be ready to provide assistance, if needed, by utilizing a planned, and immediately available, communications means (radio, hand signals, whistle, etc.). The attendant should never enter the confined space in an attempt to rescue workers until additional rescue team personnel have arrived.

CONSTRUCTION MASONRY (1926.706)

A limited access zone is to be established whenever a masonry wall is being constructed. The limited access zone shall conform to the following:

1. The limited access zone is established prior to the start of construction of the wall.

2. The limited access zone is equal to the height of the wall to be reconstructed plus four feet, and must run the entire length of the wall.

3. The limited access zone is established on the side of the wall which will be unscaffolded.

4. The limited access zone is restricted to entry by employees actively engaged in constructing the wall; no other employees are permitted to enter the zone.

5. The limited access zone remains in place until the wall is adequately supported to prevent overturning or collapse, unless the height of wall is over eight feet, in which case the limited access zone shall remain in place until the requirements of paragraph (b) of this section have been met.

All masonry walls over eight feet in height are to be adequately braced to prevent overturning or collapse unless the wall is adequately supported so that it will not overturn or collapse. The bracing shall remain in place until permanent supporting elements of the structure are in place.

CONVEYORS (1926.555)

Since conveyors present a moving hazard to workers, great care is to be taken to assure that operators have access to the on/off switch and an audible warning exists prior to the conveyor starting.

Workers are prohibited from riding moving chains, conveyors, or similar equipment. Many workers have been caught between the belt and rollers which has resulted in loss of fingers, hands, arms, or other severe injuries and even death.

Figure 19. Example of conveyor hazards with some guarding

Thus, conveyors and other such equipment need to be equipped with emergency cut-off switches, and workers should be protected from the conveyors' moving parts by guards. All nip points and noise or tail pieces should be protected so workers cannot come into contact with moving pulleys, gears, belts, or rollers (see Figure 19). The conveyor should be shut down and locked out/tagout during maintenance and repair work.

If material can fall from the conveyor, overhead protection needs to be in place, or the walkway under the conveyor should have barriers or barricades to prevent individuals from passing through. For specific questions, consult ANSI B20.1-1957.

Figure 20. Crane with outriggers deployed and warning barriers

CRANES AND DERRICKS (1926.550)

Rated Loads

The employer shall comply with the manufacturer's specifications and limitations applicable to the operation of any and all cranes and derricks. Where manufacturer's specifications are not available, the limitations assigned to the equipment are based on the determinations of a qualified engineer competent in this field, and such determinations will be appropriately documented and recorded. Attachments used with cranes must not exceed the capacity, rating, or scope recommended by the manufacturer.

Rated load capacities, recommended operating speeds, special hazard warnings, or instructions are to be conspicuously posted on all equipment. Instructions or warnings must be visible to the operator while he/she is at his/her control station. All employees are kept clear of suspended loads or loads about to be lifted.

Cranes must deploy outriggers in order to widen the base to be able to lift their intended loads. Warning barriers are erected to keep workers out of the area or operation and the swing radius of the crane itself (see Figure 20).

Hand Signals

Hand signals to crane and derrick operators are those prescribed by the applicable ANSI standard for the type of crane in use. An illustration of the signals must be posted at the jobsite. (See Figure 21.)

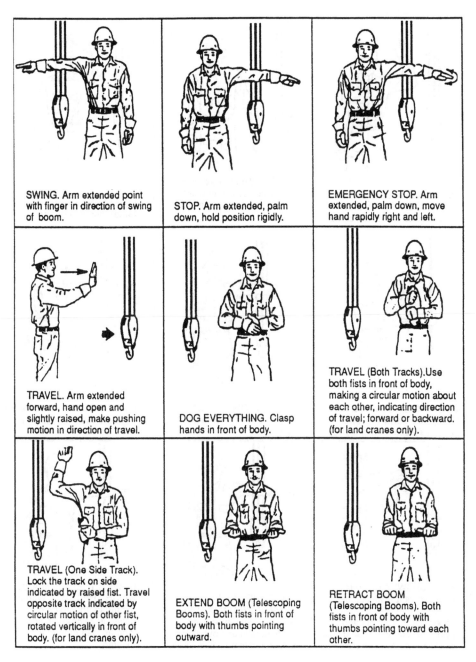

Figure 21.   Hand signals to be used with cranes and derricks. Courtesy of Department of Energy

Crane Inspections

The employer must designate a competent person who inspects all machinery and equipment prior to each use, and during use, to make sure it is in safe operating condition. (See Figure 22 for an example of a crane inspection form.) Any deficiencies are repaired, or

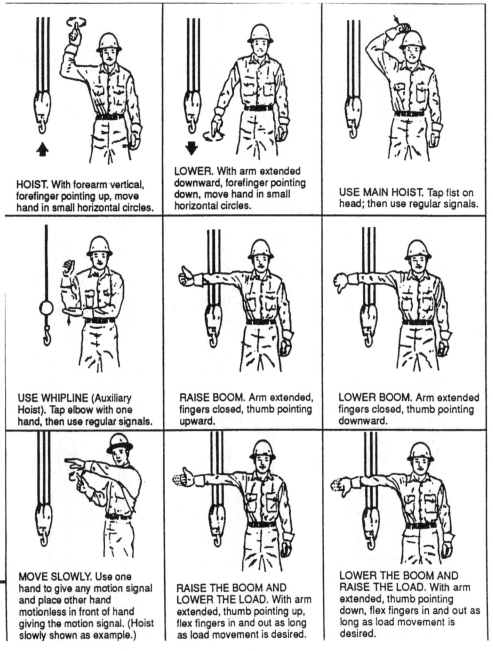

Figure 21.   Hand signals to be used with cranes and derricks. Courtesy of Department of Energy (continued)

defective parts replaced, before continued use. A thorough annual inspection of the hoisting machinery is made by a competent person, or by a government or private agency recognized by the U.S. Department of Labor. The employer maintains a record of the dates and results of inspections for each hoisting machine and piece of equipment.

MONTHLY CRANE INSPECTION FORM

Make: _____ Model: _____ Serial # _____

Date of Inspection: _____ Crane Location: _____

Job # _____ Inspected by: _____

AREA	OK	R/N	N/A	R/C
General:				
Appearance				
Paint				
Glass				
Cab				
Fire Extinguisher (5BC)				
Manuals (Parts & Operators)				
Load Charts				
Hand Signals				
Grease/Oil				
Boom				
Angle Indicator				
Warning Signs				
Steps				
Guards in Place				
Hands and Grab Rails				
Access to Roof				
Anti-Lock Block				
Back-up Alarm				
Engine:				
Operating Condition				
Oil Level and Condition				
Hour Meter				
Engine Instruments				
Cooling System				
Anti-Freeze				
Battery Condition				
Hose Condition				
Clamps				
Air System				
Pressure				
Belt Condition				
Filter				
Compressor				
Converters				
Engine Clutch				
Day Tank (Converters)				
Hydraulic Reservoir				
Oil Pressure				
Operating Temperature				
Transmission Case				
Electrical & Control System:				
Electrical Components & Assemblies				
Emergency Stop Switch				
Overload Switch				
Master Switches or Drum Controller				
Switches, Contacts & Relays				
Main Control Station				
Remote Control Station & Bucket/Ground Controls				

AREA	OK	R/N	N/A	R/C
Tracks:				
Chains				
Sprockets				
Idlers				
Pins				
Track Adjustments				
Roller Path				
Travel Brake				
Carrier:				
Tire Condition				
Brakes				
Steering				
Outriggers & Pads				
Glass				
Controls				
Fire Extinguisher (5BC)				
Horn				
Turn Signals				
Lights				
Transmission				
Frame				
Battery				
License No. (Current)				
Boom:				
No. or Type				
Length				
Tagline				
Swing System				
Clutch				
Brake				
House Rollers				
Hook Rollers				
Swing Gears				
Drum Shaft				
Clutches				
Main Line				
Aux. Line				
Brakes				
Main Line				
Aux. Line				
Boom:				
Boom Hoist				
Worn Gear				
Brass Gear				
Brake				
Clutches				
Hydraulic System				
Pawl				
Lubrication				
Guard in Place				

Figure 22. Crane inspection form

MONTHLY CRANE INSPECTION FORM (CONTINUED)

AREA	OK	R/N	N/A	R/C	AREA	OK	R/N	N/A	R/C
Indicators: Levels: Boom Angle and Length Drum Rotation: Load					**Cables & Wire Rope:** Boom Hoist Load Line Aux./Whip Line Ringer Boom Line Ringer Aux./Whip Line Load Line Wedge Socket Aux./Whip Line Wedge Socket Cable Clamps Kinks & Broken Strands (Any Abuse) Jib Pendants				
Boom & Attachments: Number/Type Boom Inventory Point Heel 10 Feet 20 Feet 30 Feet 40 Feet 50 Feet Load Block (Capacity) Hook Safety Latch Boom Pins/Cotter Pins Gantry Equalizer Cable Rollers Cable Guides Spreader Bar Jib (Type & Length) Lubrication Jib Inventory Point 10 Feet 15 Feet 20 Feet Headache Ball (Capacity) Cable Rollers Pendant Lines Off-Set Links Boom Stops Auto Boom Stop Counterweights					**Records:** Current Certification Posting Operator Instructions Preventive Maintenance Properly Marked Operator Controls, Levels & Diagrams Info. & Warning Decals				

OK - Satisfactory   R/N - Repairs Needed   N/A - Not Applicable   R/C - Repairs Completed

Figure 22. Crane inspection form (continued)

Wire Rope

Wire rope safety factors are in accordance with American National Standards Institute B 30.5-1968 or SAE J959-1966. Wire rope is taken out of service when any of the following conditions exist:

1. In running ropes, six randomly distributed broken wires in one lay, or three broken wires in one strand in one lay.

2. Wear of one-third of the original diameter of outside individual wires. Kinking, crushing, bird caging, or any other damage resulting in distortion of the rope structure.

3. Evidence of any heat damage from any cause.

4. Reductions from nominal diameter of more than one-sixty-fourth inch for diameters up to, and including five-sixteenths inch; one-thirty-second inch for diameters three-eighths inch to, and including one-half inch; three-sixty-fourths inch for diameters nine-sixteenths inch to, and including three-fourths inch; one-sixteenth inch for diameters seven-eighths inch to one and one-eighth inch

inclusive; three-thirty-seconds inch for diameters one and one-fourth to one and one-half inch inclusive.

5. In standing ropes, more than two broken wires in one lay in sections beyond end connections, or more than one broken wire at an end connection.Guarding

Belts, gears, shafts, pulleys, sprockets, spindles, drums, fly wheels, chains, or other reciprocating, rotating, or moving parts or equipment are guarded if such parts are exposed to contact by employees, or otherwise create a hazard. Guarding must meet the requirements of the American National Standards Institute B 15.1-1958 Rev., Safety Code for Mechanical Power Transmission Apparatus.

Accessible areas within the swing radius of the rear of the rotating superstructure of the crane, either permanently or temporarily mounted, are barricaded in such a manner as to prevent an employee from being struck or crushed by the crane.

All exhaust pipes are guarded or insulated in areas where contact by employees is possible in the performance of normal duties. Whenever internal combustion engine-powered equipment exhausts in enclosed spaces, tests are made and recorded to see that employees are not exposed to unsafe concentrations of toxic gases or oxygen-deficient atmospheres.

All windows in cabs are of safety glass, or equivalent, that introduces no visible distortion that will interfere with the safe operation of the machine.

Where necessary for rigging or service requirements, a ladder or steps are provided to give access to a cab roof. Guardrails, handholds, and steps are provided on cranes for easy access to the car and cab, conforming to American National Standards Institute B30.5. Platforms and walkways are to have anti-skid surfaces.

Fueling

Fuel tank filler pipe is located in such a position, or protected in such manner, as to not allow spill or overflow to run onto the engine, exhaust, or electrical equipment of any machine being fueled. An accessible fire extinguisher of 5BC rating, or higher, is available at all operator stations or cabs of equipment. All fuels are transported, stored, and handled to meet the rules of the construction fire safety regulation. When fuel is transported by vehicles on public highways, Department of Transportation rules contained in 49 CFR Parts 177 and 393 concerning such vehicular transportation are considered applicable.

Electrical Concerns

Except where electrical distribution and transmission lines have been deenergized and visibly grounded at the point of work, or where insulating barriers, not a part of, or an attachment to the equipment or machinery, have been erected to prevent physical contact with the lines, equipment, or machines, are to be operated proximate to power lines only in accordance with the following:

1. For lines rated 50 kV. or below, minimum clearance between the lines and any part of the crane or load shall be 10 feet.

2. For lines rated over 50 kV., minimum clearance between the lines and any part of the crane or load shall be 10 feet plus 0.4 inch for each 1 kV. over 50 kV., or twice the length of the line insulator, but never less than 10 feet.

3. In transit with no load and boom lowered, the equipment clearance is a minimum of 4 feet for voltages less than 50 kV., and 10 feet for voltages over 50 kV., up to and including 345 kV., and 16 feet for voltages up to and including 750 kV.

4. A person is designated to observe clearance of the equipment and give timely warning for all operations where it is difficult for the operator to maintain the desired clearance by visual means.

5. Cage-type boom guards, insulating links, or proximity warning devices may be used on cranes, but the use of such devices does not alter the requirements of any other regulation of this part, even if such device is required by law or regulation.

6. Any overhead wire is considered to be an energized line, unless and until the person owning such line or the electrical utility authorities indicate that it is not an energized line and it has been visibly grounded.

7. Prior to work near transmitter towers, where an electrical charge can be induced in the equipment or materials being handled, the transmitter is deenergized or tests are made to determine if electrical charge is induced on the crane. The following precautions are taken, when necessary, to dissipate induced voltage: the equipment is provided with an electrical ground directly to the upper rotating structure supporting the boom, and ground jumper cables are attached to materials being handled by the boom equipment when an electrical charge is induced while working near energized transmitters. Crews are provided with nonconductive poles having large alligator clips, or other similar protection to attach the ground cable to the load. Combustible and flammable materials are removed from the immediate area prior to operations.

Modifications

No modifications or additions, which affect the capacity or safe operation of the equipment, are made by the employer without the manufacturer's written approval. If such modifications or changes are made, the capacity, operation, and maintenance instruction plates, tags, or decals are changed accordingly. In no case is the original safety factor of the equipment to be reduced. The employer shall comply with Power Crane and Shovel Association Mobile Hydraulic Crane Standard No. 2. Sideboom cranes, mounted on wheel or crawler tractors, must meet the requirements of SAE J743a-1964.

Crawler, Locomotive, and Truck Cranes

Crawler, locomotive, and truck cranes must have positive stops on all jibs to prevent their movement of more than 5 degrees above the straight line of the jib and boom on conventional type crane booms. The use of cable type belly slings does not constitute compliance with this rule. All crawler, truck, or locomotive cranes in use must meet the applicable requirements for design, inspection, construction, testing, maintenance, and operation as prescribed in the ANSI B30.5-1968, Safety Code for Crawler, Locomotive and Truck Cranes. However, the written, dated, and signed inspection reports and records of the monthly inspection of critical items prescribed in Section 5-2.1.5 of the ANSI B30.5-1968 standard are not required.

Instead, the employer shall prepare a certification record which includes the date the crane items were inspected, the signature of the person who inspected the crane items, and a serial number, or other identifier, for the crane inspected. The most recent certification record shall be maintained on file until a new one is prepared.

Hammerhead Tower Cranes

For hammerhead tower cranes, adequate clearance is to be maintained between moving and rotating structures of the crane and fixed objects, in order to allow for safe passage of employees. Each employee required to perform duties on the horizontal boom of the hammerhead tower cranes is protected against falling by guardrails, or by a personal fall arrest system. Buffers are provided at both ends of travel of the trolley. Cranes mounted on rail tracks are equipped with limit switches limiting the travel of the crane on the track, and stops or buffers at each end of the tracks. All hammerhead tower cranes in use are required to meet the applicable requirements for design, construction, installation, testing, maintenance, inspection, and operation as prescribed by the manufacturer.

Overhead and Gantry Cranes

For overhead and gantry cranes, the rated load of the crane is plainly marked on each side of the crane, and if the crane has more than one hoisting unit, each hoist has its rated load marked on it or its load block; this marking is clearly legible from the ground or floor. Bridge trucks are equipped with sweeps which extend below the top of the rail and project in front of the truck wheels. Except for floor-operated cranes, a gong or other effective audible warning signal is provided for each crane equipped with a power traveling mechanism. All overhead and gantry cranes in use meet the applicable requirements for design, construction, installation, testing, maintenance, inspection, and operation as prescribed in the ANSI B30.2.0-1967, Safety Code for Overhead and Gantry Cranes.

Derricks

All derricks in use meet the applicable requirements for design, construction, installation, inspection, testing, maintenance, and operation as prescribed in American National Standards Institute B30.6-1969, Safety Code for Derricks. For floating cranes and derricks, when a mobile crane is mounted on a barge, the rated load of the crane does not exceed the original capacity specified by the manufacturer.

A load rating chart, with clearly legible letters and figures, is provided with each crane, and securely fixed at a location easily visible to the operator. When load ratings are reduced to stay within the limits listed for a barge with a crane mounted on it, a new load rating chart is to be provided.

Floating Cranes and Derricks

Mobile cranes on barges must be positively secured. For permanently mounted floating cranes and derricks, when cranes and derricks are permanently installed on a barge, the capacity and limitations of use are based on competent design criteria. A load rating chart, with clearly legible letters and figures, is provided and securely fixed at a location easily visible to the operator. Floating cranes and floating derricks in use must meet the applicable requirements for design, construction, installation, testing, maintenance, and operation as prescribed by the manufacturer. The employer shall comply with the applicable requirements for protection of employees working onboard marine vessels.

Crane- and Derrick-Suspended Personnel Platforms

The use of a crane or derrick to hoist employees on a personnel platform is prohibited, except when the erection, use, and dismantling of conventional means of reaching the worksite, such as a personnel hoist, ladder, stairway, aerial lift, elevating work platform, or scaffold would be more hazardous, or is not possible because of structural design or worksite conditions. Hoisting of the personnel platform is performed in a slow, controlled, cautious manner with no sudden movements of the crane, derrick, or the platform.

Platform Operations

Load lines must be capable of supporting, without failure, at least seven times the maximum intended load, except where rotation resistant rope is used; when they are used, the lines must be capable of supporting, without failure, at least ten times the maximum intended load. The required design factor is achieved by taking the current safety factor of 3.5 and applying the 50 percent derating of the crane capacity. The total weight of the loaded personnel platform and related rigging does not exceed 50 percent of the rated capacity for the radius and configuration of the crane or derrick. Load and boom hoist drum brakes, swing brakes, and locking devices, such as pawls or dogs, are to be engaged when the occupied personnel platform is in a stationary position.

The use of machines having live booms (booms in which lowering is controlled by a brake without aid from other devices which slow the lowering speeds) is prohibited. Cranes and derricks with variable angle booms are equipped with a boom angle indicator readily visible to the operator. Cranes with telescoping booms are equipped with a device to indicate clearly to the operator, at all times, the boom's extended length or an accurate determination of the load radius, which will be used during the lift, is made prior to hoisting personnel.

The crane must be uniformly level within one percent of level grade and located on firm footing. Cranes equipped with outriggers must be fully deployed, following the manufacturer's specifications, insofar as applicable, when hoisting employees.

A positive acting device is used which prevents contact between the load block or overhaul ball and the boom tip (anti-two-blocking device), or a system is used which deactivates the hoisting action before damage occurs in the event of a two-blocking situation (two-block damage prevention feature).

The load line hoist drum has a system or device on the power train, other than the load hoist brake, which regulates the lowering rate of speed of the hoist mechanism (controlled load lowering); free fall is prohibited.

Platform Specifications

The personnel platform and suspension system shall be designed by a qualified engineer or a qualified person competent in structural design. The suspension system is designed to minimize tipping of the platform due to movement of employees occupying the platform. The personnel platform, except for the guardrail and personnel fall arrest system anchorages, must be capable of supporting, without failure, its own weight and at least five times the maximum intended load. The criteria for guardrail systems and personal fall arrest system anchorages are described under fall protection. Each personnel platform is equipped with a guardrail system and is enclosed at least from the toeboard to the mid-rail, with either solid construction or expanded metal, having openings no greater than 1/2 inch. A grab rail is to be installed inside the entire perimeter of the personnel platform. Access gates, if installed, are not to swing outward during hoisting. All access gates, including sliding or folding gates, are equipped with a restraining device to prevent accidental opening.

Headroom is to be provided which allows employees to stand upright on the platform. In addition to the use of hard hats, employees are protected by overhead protection on the personnel platform when employees are exposed to falling objects.

All rough edges exposed to contact by employees are to be surfaced or smoothed in order to prevent puncture or laceration injuries to employees. All welding of the personnel platform, and its components, must be performed by a qualified welder who is familiar with the weld grades, types, and material specified in the platform design.

The personnel platform is to be conspicuously posted with a plate, or other permanent markings, which indicate the weight of the platform and its rated load capacity or maximum intended load. The personnel platform must not be loaded in excess of its rated load capacity. When a personnel platform does not have a rated load capacity, then the personnel platform is not loaded in excess of its maximum intended load.

The number of employees occupying the personnel platform cannot exceed the number required for the work being performed. Personnel platforms are used only for employees, their tools, and the materials necessary to do their work; when not hoisting personnel, they are not to be used to hoist only materials or tools.

Materials and tools used during a personnel lift must be secured, to prevent displacement, and evenly distributed within the confines of the platform while the platform is suspended.

Rigging

When a wire rope bridle is used to connect the personnel platform to the load line, each bridle leg is connected to a master link or shackle in such a manner to ensure that the load is evenly divided among the bridle legs. Hooks on overhaul ball assemblies, lower load blocks, or other attachment assemblies, shall be of a type that can be closed and locked, eliminating the hook throat opening. Alternatively, an alloy anchor type shackle with a bolt, nut, and retaining pin may be used.

Wire rope, shackles, rings, master links, and other rigging hardware must be capable of supporting, without failure, at least five times the maximum intended load applied or transmitted to that component. Where rotation resistant rope is used, the slings are capable of supporting, without failure, at least ten times the maximum intended load. All eyes in wire rope slings are fabricated with thimbles.

Bridles and associated rigging for attaching the personnel platform to the hoist line are only used for the platform and the necessary employees, their tools, and the materials necessary to do their work; they are not used for any other purpose when not hoisting personnel.

Inspection and Proof Test

A trial lift, with the unoccupied personnel platform loaded at least to the anticipated lift weight, is made from ground level, or any other location where employees will enter the platform to each location at which the personnel platform is to be hoisted and positioned. This trial lift is performed immediately prior to placing personnel on the platform. The operator determines that all systems, controls, and safety devices are activated and functioning properly, that no interferences exist, and that all configurations necessary to reach those work locations, will allow the operator to remain under the 50 percent limit of the hoist's rated capacity.

Materials and tools to be used during the actual lift can be loaded in the platform for the trial lift. A single trial lift may be performed at one time for all locations that are to be reached from a single setup position.

The trial lift is repeated prior to hoisting employees whenever the crane or derrick is moved and set up in a new location, or returned to a previously used location. Additionally, the trial lift is repeated when the lift route is changed, unless the operator determines that the route change is not significant (i.e., the route change would not affect the safety of hoisted employees). After the trial lift and just prior to hoisting personnel, the platform is hoisted a few inches and inspected to ensure that it is secure and properly balanced. Employees are not hoisted unless the following conditions are determined to exist:

1.   Hoist ropes are free of kinks.

2.   Multiple part lines are not twisted around each other.

3.   The primary attachment is centered over the platform.

4.   The hoisting system is inspected, if the load rope is slack, to ensure all ropes are properly stated on drums and in sheaves.

A visual inspection of the crane or derrick, rigging, personnel platform, and the crane or derrick base support, or ground, must be conducted by a competent person immediately after the trial lift. This must be done to determine whether the testing has exposed any defect or produced any adverse effect upon any component or structure. Any defects found during inspections, which create a safety hazard, are to be corrected before hoisting personnel.

At each jobsite, prior to hoisting employees on the personnel platform and after any repair or modification, the platform and rigging are proof tested to 125 percent of the platform's rated capacity by holding it in a suspended position for five minutes with the test load evenly distributed on the platform. (This may be done concurrently with the trial lift.) After proof testing, a competent person must inspect the platform and rigging. Any deficiencies found are to be corrected and another proof test conducted. Personnel hoisting is not conducted until the proof testing requirements are satisfied.

<u>Work Practices</u>

Workers are to keep all parts of the body inside the platform during raising, lowering, and positioning. This provision does not apply to an occupant of the platform performing the duties of a signal person. Before employees exit or enter a hoisted personnel platform that has not landed, the platform is to be secured to the structure where the work is to be performed, unless securing to the structure creates an unsafe situation. Tag lines must be used unless their use creates an unsafe condition. The crane or derrick operator is to remain at the controls at all times when the crane engine is running and the platform is occupied. Hoisting of employees is promptly discontinued upon indication of any dangerous weather conditions or other impending danger.

Employees being hoisted must remain in continuous sight of, and in direct communication with, the operator or signal person. In those situations where direct visual contact with the operator is not possible, and the use of a signal person would create a greater hazard for the person, direct communication, such as by radio, may be used.

Except over water, employees occupying the personnel platform must use a body harness system with lanyard appropriately attached to the lower load block or overhaul ball, or attached to a structural member within the personnel platform which is capable of supporting an employee who might fall while using the anchorage. No lifts are made on another of the crane's or derrick's loadlines while personnel are suspended on a platform.

Traveling

Hoisting of employees while the crane is traveling is prohibited, except for portal, tower, and locomotive cranes, or where the employer demonstrates that there is no less hazardous way to perform the work. Under any circumstances where a crane would travel while hoisting personnel, the employer must implement the following procedures to safeguard employees:

1.   Crane travel is restricted to a fixed track or runway.

2.   Travel is limited to the load radius of the boom used during the lift.

3.   The boom must be parallel to the direction of travel.

A complete trial run is performed to test the route of travel before employees are allowed to occupy the platform. This trial run can be performed at the same time as the trial lift which tests the route of the lift. If travel is done with a rubber tired-carrier, the condition and air pressure of the tires must be checked. The chart capacity for lifts on rubber is to be used for application of the 50 percent reduction of rated capacity. Outriggers may be partially retracted, as necessary, for travel.

Prelift Meeting

A meeting, attended by the crane or derrick operator, signal person(s) (if necessary for the lift), employee(s) to be lifted, and the person responsible for the task to be performed, is held to review the appropriate requirements and procedures to be followed. This meeting is held prior to the trial lift at each new work location, and is repeated for any employees newly assigned to the operation.

Accidents involving cranes are often caused by human actions or failure to act. Therefore, the company must employ competent operators who are physically and mentally fit and thoroughly trained in the safe operation of crane and rigging equipment, and the safe handling of loads. Upon employment, the crane operator must be assigned to work with the crane and rigging foreman, and the new operator should work only on selective jobs which will be closely monitored for a period of not less than one week.

DEMOLITION (1926.850)

When undertaking demolition work, an engineering survey must be completed prior to the start of the demolition project to determine condition of the framing, floors, and walls. Once work is ready to start, all utilities or energy sources will be turned off, all floors or walls to be demolished will be shored or braced, and all wall or floor openings must be protected. If hazardous chemicals, gases, explosives, flammable materials, or similarly dangerous substances have been used in pipes, tanks, or other equipment on the property, testing and purging must be performed to eliminate the hazard prior to demolition. During demolition involving combustible materials, charged hose lines, supplied by hydrants, water tank trucks with pumps, or equivalent, are made available. During demolition or alterations, existing automatic sprinkler installations are retained in service as long as reasonable. The operation of sprinkler control valves are permitted only by properly authorized persons. Modification of sprinkler systems, to permit alterations or additional demolition, should be expedited so that the automatic protection may be returned to service as quickly as possible. Sprinkler control valves are checked daily at close of work to ascertain that the protection is in service.

Only use stairways, passageways, and ladders which are designated as a means of access to the structure of a building. Stairs, passageways, ladders, and incidental equipment must be periodically inspected and maintained in a clean and safe condition. Stairwells must be properly illuminated and completely and substantially covered over at a point not less than two floors below the floor on which work is being performed.

When balling or clamming is being performed, never enter any area which may be adversely affected by demolition operations unless you are needed to perform these operations. During demolition, a competent person must make continued inspections as the work progresses to detect any hazards resulting from weakened or deteriorated floors, walls, or loosened material.

Chutes (1926.852)

No material is dropped outside the exterior wall of the structure unless the area is effectively protected. All chutes or sections thereof which are at an angle of more then 45 degrees from horizontal, are entirely enclosed, except for openings equipped with closures or about floor level for insertion of materials. Openings must not exceed 48 inches in height (measured along the wall of the chute). All stories below the top floor opening are to be kept closed except when in use. A substantial gate is to be installed in each chute near the discharge end. A competent person must be assigned to control the gate and the backing and loading of trucks. Any chute opening into which debris is dumped shall be protected by a guardrail 42 inches above the floor or surface on which workers stand to dump materials. Any space between the chute and edge of opening around the chute shall be covered. When material is dumped from mechanical equipment or a wheelbarrow, a toeboard or bumper, not less than four inches thick and six inches high is to be provided around the chute opening. Chutes must be strong enough to withstand the impact of materials dumped into it.

Removal of Materials through Floor Openings (1926.853)

Any openings in the floor are to be less than 25% of the aggregate of the total floor unless lateral supports of the floor remain in place. Any weakened floor is to be shored to support demolition loads.

Removal of Walls, Masonry Sections, and Chimneys (1926.854)

Masonry walls, or other masonry sections, are not permitted to fall upon the floors of buildings in such amount as to exceed the load limit of the floor. No wall section which is more than one story in height is allowed to stand without lateral bracing, unless originally designed to do so, and still has that integrity. No unstable wall shall exist at the end of a shift. Workers are not to work on top of walls when weather constitutes a hazard. No structural or lead-supporting member are to be cut or removed until all stories above have been demolished or removed, except for disposal openings or equipment installation. Floor openings within 10 feet of any wall being demolished must be planked solid or workers must be kept out of the area below.

In skeletal-steel constructed buildings, the steel frame may be left in place during the demolition of masonry, but all beams, girders, or other structural supports are to be kept clear of debris and loose materials. Walkways and ladders must be provided so workers can safely reach or leave any scaffold or wall. Walls which retain or support earth or adjoining structures are not removed until the earth or adjoining structure is supported.

Walls against which debris is piled must be capable of supporting that load.

Manual Removal of Floors (1926.855)

Openings cut into a floor must extend the full span of the arch between supports. Before floor arches are removed, debris and other material are to be removed. Two by ten planks must be full size, undressed, and have open spaces no more than 16 inches in order for workers to safely stand on them during the break down of floor arches between beams. Walkways not less than 18 inches wide, formed of wooden planks no less than 2 inches thick, or equivalent strength, if metal, are to be provided to workers to prevent them from walking on exposed beams to reach work points.

Stringers of ample strength must be installed to support flooring planks. These stringers are supported by floor beams or girders and not floor arches. Planks are to be laid together over solid bearings with ends overlapped at least one foot. Employees are not allowed in the area directly below where floor arches are being removed. This area is barricaded to prevent access. Floor arches are removed only after the surrounding area for a distance of 20 feet has been cleared of debris and other unnecessary materials.

Removal of Walls, Floors, and Material with Equipment (1926.856)

All floors and work surfaces are to be of sufficient strength to support any mechanical equipment being used. Curbs or stop-logs are put in place to prevent equipment from running over the edges.

Storage (1926.857)

The allowable floor loads are not to be exceeded by the waste or debris being stored there. Floor boards may be removed in buildings having wooden floor construction not more than one floor above grade. This will provide storage space for debris and assure that no falling material will endanger the stability of the structure. Wood floor beams, which brace interior walls or support free-standing exterior walls, must be left in place until equivalent support can be installed. Storage space, into which materials are dumped, is blocked off, except for the openings necessary for material removal. These openings are to be kept closed when not in use.

Removal of Steel Construction (1926.858)

Planking shall be used when floor arches have been removed and workers are engaged in razing the steel frame. Cranes, derricks, and other hoisting equipment must meet the construction specifications.

Steel construction is to be dismantled column length by column length, and tier by tier (columns may be in two-story lengths). Structural members being dismembered are not to be overstressed.

Mechanical Demolition (1926.859)

Workers are not permitted in areas where balling and clamming is being performed. Only essential workers are permitted in those areas at any other time. The weight of the demolition ball shall not exceed 50% of the crane's rated load, based on the length of the boom and the maximum angle of operation, and shall not exceed 25% on the nominal break strength of the line by which it is suspended, whichever is the lesser value. The crane load line is to be kept as short as possible. The ball must be attached to the load line with a swivel-type connection to prevent twisting, and it must be attached in such a manner that it cannot accidentally become disconnected. All steel members of walls must be cut free prior to pulling over a wall or portion thereof. Roof cornices or other ornamental stonework have to be removed prior to pulling over walls. To detect hazards resulting from weakened or loosened materials, a competent person must conduct inspections as demolition progresses. Workers are not permitted to work where such hazards exist until corrective action is taken.

Selective Demolition by Explosives (1926.860)

Demolition, using explosives, is conducted following the specifications of Subpart U.

DISPOSAL CHUTES (1926.252)

Disposal chutes are installed to prevent the hazards which exist from materials, etc., dropped from heights greater than 20 feet (see Figure 23). When materials are dropped to the exterior of a building, the chutes must be enclosed. The disposal chute should mitigate the danger from overhead hazard from falling materials.

At times materials or debris are dropped through holes in the floor. The landing area for this type of waste must have a 42 inch high barricade which is six feet back from the edge on all sides of the floor opening's edges. Also, warnings are to be posted at each level denoting the fall materials hazard.

DIVING (1926.1071)

Most construction companies do not conduct diving operations, but, of course, there are those who specialize in this type of operation and, thus, must comply with 29 CFR 1926 Subpart Y. This is the same standard as used for commercial diving in general industry, and is found in 29 CFR 1910 Subpart T. This standard applies to diving and related support opera-≈tions conducted in connection with all types of work and employments, including general

Figure 23. Enclosed disposal chute on exterior of building

industry, construction, ship repairing, shipbuilding, shipbreaking, construction, and longshoring.

Each dive team member must have the experience or training necessary to perform the assigned tasks in a safe and healthful manner. Each dive team member shall have experience or training in the following:

1. The use of tools, equipment, and systems relevant to assigned tasks.

2. Techniques of the assigned diving mode.

3. Diving operations and emergency procedures.

All dive team members are to be trained in cardiopulmonary resuscitation and first aid (American Red Cross standard course or equivalent). Dive team members who are exposed to, or control the exposure of others to hyperbaric conditions, are to be trained in diving-related physics and physiology. Each dive team member is assigned tasks in accordance with the employee's experience or training, except that limited additional tasks may be assigned to an employee undergoing training provided that these tasks are performed under the direct supervision of an experienced dive team member.

The employer does not require a dive team member be exposed to hyperbaric conditions against the employee's will, except when necessary to complete decompression or treatment procedures. The employer does not permit a dive team member to dive or be otherwise exposed to hyperbaric conditions for the duration of any temporary physical impairment or condition which is known to the employer and is likely to adversely affect the safety or health of a dive team member. The employer, or an employee designated by the employer, is to be at the dive location and in charge of all aspects of the diving operation affecting the safety and health of dive team members. The designated person-in-charge must have experience and training in the conduct of the assigned diving operation.

The employer develops and maintains a safe practices manual which is made available to each dive member at the dive location. The safe practices manual contains a copy of this standard and the employer's policies for implementing the requirements of the diving standard. The safe practices for each type of diving is included in the manual as follows:

1. Safety procedures and checklists for diving operations.

2. Assignments and responsibilities of the dive team members.

3. Equipment procedures and checklists.

4. Emergency procedures for fire, equipment failure, adverse environmental conditions, medical illness, and injury.

Prior to each diving operation, a list of telephone or call numbers shall be made and kept at the following dive locations:

1. An operational decompression chamber (if not at the dive location).

2. Accessible hospitals.

3. Available physicians.

4. Available means of transportation.

5. The nearest U.S. Coast Guard Rescue Coordination Center.

6. A first aid kit, appropriate for the diving operation and approved by a physician, must be available at the dive location. When used in a decompression chamber or bell, the first aid kit must be suitable for use under hyperbaric conditions. In addition to any other first aid supplies, an American Red Cross standard first aid handbook or equivalent, and a bag-type manual resuscitator with transparent mask and tubing, is to be available.

When planning a diving operation, a safety and health assessment must be made and include the following:

1. Diving mode.

2. Surface and underwater conditions and hazards.

3. Breathing gas supply (including reserves).

4. Thermal protection.

5. Diving equipment and systems.

6. Dive team assignments and physical fitness of dive team members (including any impairment known to the employer).

7. Repetitive dive designation or residual inert gas status of dive team members.

8. Decompression and treatment procedures (including altitude corrections).

9. Emergency procedures.

To minimize hazards to the dive team, diving operations are coordinated with other activities in the vicinity which are likely to interfere with the diving operation. Dive team members are briefed on the tasks to be undertaken, safety procedures for the diving mode, any unusual hazards or environmental conditions likely to affect the safety of the diving operation, and any modifications to operating procedures necessitated by the specific diving operation.

Prior to making individual dive team member assignments, the employer must inquire into the dive team member's current state of physical fitness, and indicate to the dive team member the procedure for reporting physical problems or adverse physiological effects during and after the dive. The breathing gas supply system, including reserve breathing gas supplies, masks, helmets, thermal protection, and bell handling mechanism (when appropriate), are to be inspected prior to each dive. When diving from surfaces other than vessels, and diving into areas capable of supporting marine traffic, a rigid replica of the international code flag "A," which is at least one meter in height, must be displayed at the dive location. It is to be displayed in such a manner that it allows all-round visibility, and must be illuminated during night diving operations.

During a dive the following requirements are applicable to each diving operation, unless otherwise specified. There must be: a means provided which is capable of supporting the diver when entering and exiting the water, and that means must extend below the water surface to enable the diver an easy exit of the water; a means to assist an injured diver from the water or into a bell; and an operational two-way voice communication system between each surface-supplied air or mixed-gas diver and a dive team member at the dive location or bell (when provided or required at the bell and dive location).

At each dive location it is expected that

1. An operational, two-way communication system is available at the dive location to obtain emergency assistance.

2. Decompression, repetitive, and no-decompression tables (as appropriate) are at the dive location.

3. Depth-time profile, including, when appropriate, any breathing gas changes, shall be maintained for each diver during the dive, including decompression.

4. Hand-held electrical tools and equipment are deenergized before being placed into, or retrieved from the water.

5. Hand-held power tools are not supplied with power from the dive location until requested by the diver.

6. A current supply switch, which interrupts the current flow to the welding or burning electrode, is to be tended by a dive team member in voice communication with the diver performing the welding or burning, and it must be kept in the open position except when the diver is welding or burning. Also, the welding machine frame is to be grounded. Welding and burning cables, electrode holders, and connections shall be capable of carrying the maximum current required by the work and must be properly insulated, and insulated gloves must be provided to divers performing welding and burning operations. Prior to welding or burning on closed compartments, structures, or pipes which contain a flammable vapor, or where a flammable vapor may be generated by the work, they shall be vented, flooded, or purged with a mixture of gases which will not support combustion.

7. Employers who transport, store, and use explosives must comply with the provisions of and 1926.912 of Title 29 of the Code of Federal Regulations. Electrical continuity of explosive circuits are not to be tested until the diver is out of the water. Explosives must not be detonated while the diver is in the water. A blaster shall conduct all blasting operations, and no shot is fired without his approval. Loading tubes and casings of dissimilar metals are not to be used because of the possibility of electric transient currents occurring from the galvanic action of the metals and water. Only water-resistant blasting caps and detonating cords are to be used for all marine blasting. Loading shall be done through a nonsparking metal loading tube when a tube is necessary. No blast is to be fired while any vessel under way is closer than 1,500 feet of the blasting area. Those on board vessels or craft moored or anchored within 1,500 feet must be notified before a blast is fired. No blast shall be fired while any swimming or diving operations are in progress in the vicinity of the blasting area. If such operations are in progress, signals and arrangements are agreed upon to assure that no blast is fired while any person is in the water. Blasting flags must be displayed. The storage and handling of explosives aboard vessels, which are used in underwater blasting operations, must be handled according to provisions outlined therein. When more than one charge is placed under water, a float device is to be attached to an element of each charge in such a manner that it will be released by the firing. Misfires are handled in accordance with the requirements of 1926.911.

The working interval of a dive shall be terminated when: a diver requests termination; a diver fails to respond correctly to communications or signals from a dive team member; communications are lost and cannot be quickly reestablished with the diver; communications between a dive team member at the dive location, and the designated person-in-charge or the person controlling the vessel in liveboating operations; or a diver begins to use diver-carried reserve breathing gas or the dive-location reserve breathing gas.

After the completion of any dive, the employer shall check the physical condition of the diver, instruct the diver to report any physical problems or adverse physiological effects, including symptoms of decompression sickness, advise the diver of the location of a decompression chamber which is ready for use, and alert the diver to the potential hazards of flying after diving.

For any dive which is outside the no-decompression limits and is deeper than 100 fsw, or uses mixed gas as a breathing mixture, the employer must instruct the diver to remain awake and in the vicinity of the decompression chamber, which is at the dive location, for at least one hour after the dive (including decompression or treatment, as appropriate). A decompression chamber that is capable of recompressing the diver, at the surface, to a minimum of 165 fsw (6 ATA), must be available at the dive location if: surface-supplied air diving reaches depths deeper than 100 fsw but is less than 220 fsw; mixed gas diving is less than 300 fsw; or, diving is outside the no-decompression limits and is less than 300 fsw.

The following information is recorded and maintained for each diving operation: names of the dive team members, including designated person-in-charge, date, time, location, diving modes used, general nature of work performed, approximate underwater and surface conditions (visibility, water temperature, and current), and maximum depth and bottom time for each diver.

For each dive outside the no-decompression limits, which is deeper than 100 fsw or uses mixed gas, the following additional information must be recorded and maintained: depth-time and breathing gas profiles, decompression table designation (including modification), and elapsed time since last pressure exposure, if less than 24 hours or repetitive dive designation for each diver.

Any time decompression sickness is suspected or symptoms are evident, the following additional information shall be recorded and maintained: description of decompression sickness symptoms (including depth and time of onset), and description and results of treatment.

Each incident of decompression sickness must be investigated and evaluated based on the recorded information, consideration of the past performance of decompression table used, and individual susceptibility. Corrective action should take place to reduce the probability of a recurrence, and an evaluation of the decompression procedure assessment, including any corrective action taken, should be written within 45 days of the incident.

Also, records are to be kept on the occurrence of any diving-related injury or illness which requires any dive team member to be hospitalized for 24 hours or more; specify the circumstances of the incident and the extent of any injuries or illnesses. Records and documents are required to be retained by the employer for the following period of time:

1. Dive team member medical records (physician's reports) (1910.411) – 5 years.

2. Safe practices manual (1910.420) – current document only.

3. Depth-time profile (1910.422) – until completion of the recording of dive, or until completion of decompression procedure assessment where there has been an incident of decompression sickness.

4. Recording of dive (1910.423) – 1 year, except 5 years where there has been an incident of decompression sickness.

5. Decompression procedure assessment evaluations (1910.423) – 5 years.

6. Equipment inspections and testing records (1910.430) – current entry or tag, or until equipment is withdrawn from service.

7. Records of hospitalizations (1910.440) – 5 years.

For specifics on diving modes such as scuba diving, surface supplied air diving, mixed-gas diving, lifeboating, and diving equipment, consult the standard 29 CFR 1926 Subpart Y.

DRINKING WATER

See potable water

EATING AND DRINKING AREAS (1926.51)

No food or drink is allowed to be consumed in the vicinity of toilets, or in areas exposed to toxic materials.

EGRESS (1926.34)

In every building or structure, exits are to be so arranged and maintained as to provide free and unobstructed egress from all areas, at all times, when occupied. No lock or fastening device shall be installed which prevents free escape from the inside of any building, except in mental, penal, or corrective institutions. In these institutions, supervisory personnel must be on duty at all times, and effective provisions must be made to remove occupants in case of fire or other emergencies. All exits must be marked by a readily visible sign, and where access to reach the exists are not immediately visible, they must, in all cases, be marked with readily visible directional signs. The means of egress is to be continually maintained and free of all obstructions or impediments, which would inhibit an easy and immediate exit of the building or structure, in case of a fire or other emergency.

ELECTRICAL (1926.400)

The electrical safety requirements are those necessary for the practical safeguarding of employees involved in construction work. These installation safety requirements are divided into four major areas: the electric equipment and installations used to provide electric power and light on jobsites; safety-related work practices which cover hazards that may arise from the use of electricity at jobsites, and hazards that may arise when employees accidentally contact energized lines, direct or indirect, that are above or below ground, or are passing through or near the jobsites; safety-related maintenance and environmental considerations; and safety requirements for special equipment.

These requirements cover the installation safety requirements for electrical equipment and the installations used to provide electric power and light at the jobsite. This includes installations, both temporary and permanent, used on the jobsite, but does not apply to existing permanent installations that were in place before the construction activity commenced. If the electrical installation is made in accordance with the National Electrical Code ANSI/NFPA 70-1984, it is deemed to be in compliance with the construction electrical safety requirements. The generation, transmission, and distribution of electric energy, including related communication, metering, control, and transformation installations, are not part of the following electrical safety requirements.

General Requirements (1926.403)

The general requirements state that all electrical conductors and equipment must be approved types. The employer must ensure that electrical equipment is free from recognized hazards that are likely to cause death or serious physical harm to employees. The safety of equipment is to be determined on the basis of the following considerations:

1. Suitability for installation and use in conformity with the electrical safety provisions. Suitability of equipment for an identified purpose may be evidenced by listing, labeling, or certification for that identified purpose.

2. Mechanical strength and durability, including parts designed to enclose and protect other equipment, and adequacy of the protection thus provided.

3. Electrical insulation, heating effects under conditions of use, and arcing effects.

4. Classification by type, size, voltage, current capacity, and specific use.

5.   Other factors which contribute to the practical safeguarding of employees using, or likely to come in contact with the equipment.

Listed, labeled, or certified equipment is to be installed and used in accordance with instructions included in the listing, labeling, or certification. Equipment intended to break current must have an interrupting rating at system voltage which is sufficient for the current that must be interrupted. Electrical equipment shall be firmly secured to the surface on which it is mounted. Wooden plugs, driven into holes in masonry, concrete, plaster, or similar materials, are not to be used.

Electrical equipment, which depends upon the natural circulation of air and convection principles for cooling of exposed surfaces, is to be installed so that room for air flow over such surfaces is not prevented by walls or by adjacent installed equipment. For equipment designed for floor mounting, clearance between top surfaces and adjacent surfaces shall be provided to dissipate rising warm air. Electrical equipment provided with ventilating openings is to be installed so that walls or other obstructions do not prevent the free circulation of air through the equipment.

If conductors are spliced or joined, it is to be done by splicing devices designed for that use, or by brazing, welding, or soldering with a fusible metal or alloy. Soldered splices are first spliced or joined so as to be mechanically and electrically secure without solder, and then soldered. All splices and joints and free ends of conductors are to be covered with an insulation equivalent to that of the conductors, or with an insulating device designed for the purpose.

Parts of electrical equipment, which in ordinary operation produce arcs, sparks, flames, or molten metal, are to be enclosed or separated and isolated from all combustible material.

No electrical equipment is to be used unless the manufacturer's name, trademark, or other descriptive marking is in place. The organization responsible for the product must be identified on the equipment and unless other markings are provided giving voltage, current, wattage, or other ratings as necessary. The markings must be of sufficient durability to withstand the environment involved.

Each disconnecting means for motors and appliances are to be legibly marked to indicate its purpose, unless located and arranged so the purpose is evident. Also, each service, feeder, and branch circuit, at its disconnecting means or overcurrent device, is to be legibly marked to indicate its purpose, unless located and arranged so the purpose is evident. These markings shall be of sufficient durability to withstand the environment involved.

Working Distances

When a working space has electrical equipment with 600 volts, nominal, or less, there must be sufficient access. The working space must be well maintained and provide adequate space around all electrical equipment in order to permit ready and safe operation and maintenance of such equipment. When there are live circuits that require examination, adjustment, servicing, or maintenance, the safe working distances found in Table 1 shall be followed. In addition to the dimensions shown, the working space is not to be less than 30 inches wide in front of the electrical equipment. Distances are measured from the live parts, if they are exposed, or from the enclosure front or opening, if the live parts are enclosed. Walls constructed of concrete, brick, or tile are considered to be grounded. Working space is not required in the back of assemblies of dead-front switchboards or motor control centers, where there are no renewable or adjustable parts, such as fuses or switches on the back, and where all connections are accessible from locations other than the back.

Table 1

Working Clearances
Courtesy of OSHA

	Minimum clear distance for nominal voltage to ground conditions[1]		
	(a) Feet	(b) Feet	(c) Feet
0-150	3	3	3
151-600	3	3 1/2	4

[1] Conditions (a), (b), and (c) are as follows: (a) Exposed live parts on one side and no live or grounded parts on the other side of the working space, or exposed live parts on both sides effectively guarded by insulating material. Insulated wire or insulated busbars operating at not over 300 volts are not considered live parts. (b) Exposed live parts on one side and grounded parts on the other side. (c) Exposed live parts on both sides of the workplace [not guarded as provided in Condition (a) with the operator between].

Working space is not used for storage. When normally enclosed live parts are exposed for inspection or servicing, the working space, if in a passageway or general open space, is guarded.

At least one entrance is to be provided to give access to the working space around electric equipment.

Where there are live parts normally exposed on the front of switchboards or motor control centers, the working space in front of such equipment is not to be less than three feet. The minimum headroom of working spaces about service equipment, switchboards, panelboards, or motor control centers is 6 feet 3 inches (see Figure 24).

Guarding Electrical Equipment

Live parts of electric equipment, operating at 50 volts or more, are to be guarded against accidental contact by using cabinets or other forms of enclosures, or by any of the following means:

1. By location in a room, vault, or similar enclosure that are accessible only to qualified persons (see Figure 25).

2. By partitions or screens so arranged that only qualified persons will have access to the space within reach of the live parts. Any openings in such partitions or screens must be so sized and located that persons are not likely to come into

accidental contact with the live parts, or bring conducting objects into contact with them.

3. By locating them on a balcony, gallery, or platform, and elevated and arranged so that it excludes unqualified persons.

4. By elevation of 8 feet or more above the floor or other working surface, and so installed as to exclude unqualified persons.

In locations where electric equipment would be exposed to physical damage, enclosures or guards are to be so arranged, and of such strength, as to prevent such damage. Entrances to rooms and other guarded locations containing exposed live parts, are to be marked with conspicuous warning signs which forbid unqualified persons to enter.

Figure 24. Electrician with adequate working clearance

Figure 25. Electrical enclosure accessible to qualified persons

Conductors Exceeding 600 Volts

    Conductors and equipment used on circuits exceeding 600 volts, nominal, must comply with all applicable previous requirements, as well as with provisions which supplement or modify those requirements, but does not apply to the supply side of those conductors.

    Electrical installations in a vault, room, closet or in an area surrounded by a wall, screen, or fence, access to which is controlled by lock and key or other equivalent means, are considered to be accessible to qualified persons only. A wall, screen, or fence less than 8 feet in height is not considered adequate to prevent access, unless it has other features that provide a degree of isolation equivalent to an 8-foot fence. The entrances to all buildings, rooms, or enclosures containing exposed live parts or exposed conductors operating at over 600 volts, nominal, are kept locked, or are kept under the observation of a qualified person at all times. Installations are to be accessible to qualified persons only. Electrical installations having exposed live parts are to be accessible to qualified persons only, and they must comply with workspace requirements.

    Sufficient workspace is to be provided and maintained around electrical equipment to permit ready and safe operation and maintenance of such equipment. Where energized parts are exposed, the minimum clear workspace is not to be less than 6 feet 6 inches high (measured vertically from the floor or platform), or less than 3 feet wide (measured parallel to the equipment). The depth is as required in Table 2. The workspace must be adequate to permit at least a 90-degree opening of doors or hinged panels.

Table 2

Minimum Depth of Clear Working Space in Front of Electric Equipment
Courtesy of OSHA

Nominal voltage to ground	Conditions[1]		
	(a)	(b)	(c)
	Feet	Feet	Feet
601 to 2,500	3	4	5
2,501 to 9,000	4	5	6
9,001 to 25,000	5	6	9
25,001 to 75 kV	6	8	10
Above 75kV	8	10	12

[1] Conditions (a), (b), and (c) are as follows: (a) Exposed live parts on one side and no live or grounded parts on the other side of the working space, or exposed live parts on both sides effectively guarded by insulating materials. Insulated wire or insulated busbars, operating at not over 300 volts, are not considered live parts. (b) Exposed live parts on one side and grounded parts on the other side. Walls constructed of concrete, brick, or tile are considered to be grounded surfaces. (c) Exposed live parts on both sides of the workspace [not guarded as provided in Condition (a)] with the operator between them.

The minimum clear working space in front of electric equipment such as switchboards, control panels, switches, circuit breakers, motor controllers, relays, and similar equipment, is not to be less than specified in Table 2 unless otherwise specified. Distances shall be measured from the live parts, if they are exposed, or from the enclosure front or opening, if the live parts are enclosed. However, working space is not required in back of such equipment as deadfront switchboards or control assemblies, where there are no renewable or adjustable parts (such as fuses or switches) on the back, and where all connections are accessible from locations other than the back. Where rear access is required to work on deenergized parts on the back of enclosed equipment, a minimum working space of 30 inches, horizontally, is provided.Installations Accessible to Unqualified Persons

Electrical installations that are open to unqualified persons are made with metal-enclosed equipment, are enclosed in a vault, or are in an area where access is controlled by a lock. Metal-enclosed switchgear, unit substations, transformers, pull boxes, connection boxes, and other similar associated equipment shall be marked with appropriate caution signs. If equipment is exposed to physical damage from vehicular traffic, guards shall be provided to prevent such damage. Ventilating or similar openings in metal-enclosed equipment are to be designed so that foreign objects inserted through these openings will be deflected from energized parts.

Lighting Outlets

The lighting outlets are to be so arranged that persons changing lamps or making repairs on the lighting system will not be endangered by live parts or other equipment. The points of control shall be so located that persons are not likely to come in contact with any live or moving part of the equipment while turning on the lights. Unguarded live parts, above working space, are to be maintained at elevations not less than specified in Table 3.

Table 3
Elevation of Unguarded Energized Parts Above Working Space
Courtesy of OSHA

Nominal voltage between phases	Minimum Elevation
601-7,500	8 feet 6 inches.
7,501-35,000	9 feet.
Over 35kV	9 feet + 0.37 inches per kV above 35kV.

At least one entrance, not less than 24 inches wide and 6 feet 6 inches high, must be provided to give access to the working space at the location of electric equipment. On switchboard and control panels exceeding 48 inches in width, there shall be one entrance at each end of such board, where practicable. Where bare energized parts at any voltage or insulated energized parts above 600 volts are located adjacent to such entrance, they are to be guarded.

Wiring Design and Protection (1926.404)

A conductor used as a grounded conductor is to be identifiable and distinguishable from all other conductors. A conductor used as an equipment grounding conductor must be identifiable and distinguishable from all other conductors. No grounded conductor shall be attached to any terminal or lead so as to reverse designated polarity. A grounding terminal or grounding-type device on a receptacle, cord connector, or attachment plug is not to be used for purposes other than grounding.

The employer must use either ground fault circuit interrupters, or an assured equipment grounding conductor program to protect employees on construction sites. These requirements are in addition to any other requirements for equipment grounding conductors.

Ground-Fault Circuit Interrupters

All 120-volt, single-phase 15- and 20-ampere receptacle outlets on construction sites, which are not a part of the permanent wiring of the building or structure and are in use by employees, must have approved ground-fault circuit interrupters for personnel protection. Receptacles on a two-wire, single-phase portable or vehicle-mounted generator rated not more than 5kV, where the circuit conductors of the generator are insulated from the generator frame and all other grounded surfaces, need not be protected with ground-fault circuit interrupters (see Figure 26).

Figure 26. Example of a ground-fault circuit interrupter connected to a drill

Assured Grounding Program

The employer must establish and implement an assured equipment grounding conductor program on construction sites. This program must cover all cord sets, receptacles which are not a part of the building or structure, and equipment connected by a cord and plug and are available for use, or used by employees (see Table 4). This program must comply with the following minimum requirements:

1.   A written description of the program, including the specific procedures adopted by the employer, is to be available at the jobsite for inspection and copying by

the Assistant Secretary and any affected employees. (See Figure 27.)

2.   The employer shall designate one or more competent persons to implement the program.

3.   Each cord set, attachment cap, plug and receptacle of cord sets, and any equipment connected by a cord and plug, except cord sets and receptacles which are fixed and not exposed to damage, is to be visually inspected, before each day's use, for external defects, such as deformed or missing pins or insulation damage, and for indications of possible internal damage. Equipment found damaged or defective is not to be used until repaired.

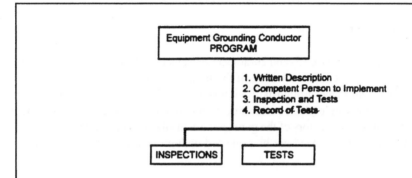

Figure 27. Assured grounding program. Courtesy of OSHA

4. The following tests must be performed on all cord sets, receptacles which are not a part of the permanent wiring of the building or structure, and cord- and plug-connected equipment required to be grounded (see Table 4).

All equipment grounding conductors are to be tested for continuity and be electrically continuous. Each receptacle and attachment cap or plug must be tested for correct attachment of the equipment grounding conductor. The equipment grounding conductor is to be connected to its proper terminal. All required tests are performed before first use, before equipment is returned to service following any repairs, before equipment is used after any incident which can be reasonably suspected to have caused damage (for example, when a cord set is run over), and at intervals not to exceed 3 months, except cord sets and receptacles which are fixed and not exposed to damage, they are to be tested at intervals not exceeding 6 months. The employer is not to make available, or permit any employee to use any equipment which has not met these requirements. Tests performed are required to be recorded. These test records shall identify each receptacle, cord set, and cord- and plug-connected equipment that passed the test, and indicate the last date it was tested, or the interval for which it was tested. This record is to be kept by means of logs, color coding, or other effective means and is to be maintained until replaced by a more current record. The record must be made available on the jobsite for inspection by the Assistant Secretary and any affected employees.

All equipment grounding conductors are to be tested for continuity and be electrically continuous. Each receptacle and attachment cap or plug must be tested for correct attachment of the equipment grounding conductor. The equipment grounding conductor is to be connected to its proper terminal. All required tests are performed before first use, before equipment is returned to service following any repairs, before equipment is used after any incident which can be reasonably suspected to have caused damage (for example, when a cord set is run over), and at intervals not to exceed 3 months, except cord sets and receptacles which are fixed and not exposed to damage, they are to be tested at intervals not exceeding 6 months. The employer

---

Table 4

Verifying Inspection

To verify inspection and testing, a piece of color coded tape will be affixed each time equipment is inspected. Four colors of tape will be used, one for each quarter of the year. The color coding system is as follows:

Color	Quarter	Expiration Date
White	First	March 31
Green	Second	June 30
Red	Third	September 30
Orange	Fourth	December 31

(Brown will be used to verify that repair is needed.)

Inspection tape will not be used for any other purpose. Storage of tape will be strictly controlled by the site superintendent. Only persons designated by the site superintendent are authorized to remove inspection tape. Unauthorized removal or defacing of inspection tape will be cause for disciplinary action.

is not to make available, or permit any employee to use any equipment which has not met these requirements. Tests performed are required to be recorded. These test records shall identify each receptacle, cord set, and cord- and plug-connected equipment that passed the test, and indicate the last date it was tested, or the interval for which it was tested. This record is to be kept by means of logs, color coding, or other effective means and is to be maintained until replaced by a more current record. The record must be made available on the jobsite for inspection by the Assistant Secretary and any affected employees.

Requirements for Outlets

Outlet devices must have an ampere rating not less than the load to be served and comply with the following:

1. A single receptacle, installed on an individual branch circuit, must have an ampere rating of not less than that of the branch circuit.

2. When connected to a branch circuit supplying two or more receptacles or outlets, receptacle ratings are to conform to set values (see Table 5).

3. The rating of an attachment plug or receptacle used for cord- and plug-connection of a motor to a branch circuit, shall not exceed 15 amperes at 125 volts, or 10 amperes at 250 volts, if individual overload protection is omitted.

Outdoor Conductors

Any branch circuit, feeder, and service conductors rated 600 volts, nominal, or less, and run outdoors as open conductors supported on poles, must provide a horizontal climbing space not less than 30 inches for power conductors below communication conductors. For power conductors alone or above communication conductors, the following must be provided: 300 volts or less–24 inches; more than 300 volts–30 inches. For communication conductors that are below power conductors and the power conductors have 300 volts or less, 24 inches must be provided, and when they have more than 300 volts, 30 inches are to be provided.

Table 5

Receptacle Ratings for Various Size Circuits
Courtesy of OSHA

Circuit rating amperes	Receptacle rating amperes
15	Not over 15
20	15 or 20
30	30
40	40 or 50
50	50

Open conductors must be at least 10 feet above finished grade, sidewalks, or from any platform or projection from which they might be reached; 12 feet over areas subject to vehicular traffic other than truck traffic; 15 feet over areas, other than those specified, that are subject to truck traffic; and 18 feet over public streets, alleys, roads, and driveways (see Figure 28).

Conductors must have a clearance of at least 3 feet from windows, doors, fire escapes, or similar locations. Conductors that run above the top level of a window are considered to be out of reach from that window and, therefore, do not have to be 3 feet away. Conductors above roof space, which are accessible to employees on foot, must have not less than 8 feet of vertical clearance from the highest point of the roof surface for insulated conductors, not less than 10 feet vertical or diagonal clearance for covered conductors, and not less than 15 feet for bare conductors, except where the roof space is also accessible to vehicular traffic, then the vertical clearance is not to be less than 18 feet. Where the roof space is not normally accessible to employees on foot, fully insulated conductors must have a vertical or diagonal clearance of not less than 3 feet.

Figure 28. Conductors over a typical construction site

Where the voltage between conductors is 300 volts or less and the roof has a slope of not less than 4 inches in 12 inches, the clearance from roofs is to be at least 3 feet. Where the voltage between conductors is 300 volts or less and the conductors do not pass over more than 4 feet of the overhang portion of the roof and they are terminated at a through-the-roof raceway or support, the clearance from roofs shall be at least 18 inches.

Lamps for outdoor lighting are to be located below all live conductors, transformers, or other electric equipment, unless such equipment is controlled by a disconnecting means that can be locked in the open position, or unless adequate clearances or other safeguards are provided for relamping operations.

Disconnects

Means are to be provided to disconnect all conductors, in a building or other structures, from the service-entrance conductors. The disconnecting means must plainly indicate whether it is in the open or closed position and must be installed at a readily accessible location nearest the point of entrance of the service-entrance conductors. Each service disconnecting means must simultaneously disconnect all ungrounded conductors.

The following requirements apply to services over 600 volts, nominal, that are service-entrance conductors and installed as open wires. They must be guarded so that no one except qualified persons has access to them. There must also be signs warning of high voltage and they are to be posted where unauthorized employees might come in contact with live parts.

Overcurrent Protection

When overcurrent protection of circuits rated 600 volts, nominal, or less is used, the conductors and equipment are to be protected from overcurrent in accordance with their ability to safely conduct current. Conductors must have sufficient ampacity to carry the load. Except for motor-running overload protection, overcurrent devices shall not interrupt the continuity of the grounded conductor, unless all conductors of the circuit are opened simultaneously. Except for devices provided for current-limiting on the supply side of the service disconnecting means, all cartridge fuses which are accessible to other than qualified persons, and all fuses and thermal cutouts on circuits over 150 volts to ground, are to be provided with a disconnecting means. These disconnecting means are to be installed so that the fuse or thermal cutout can be disconnected from its supply without disrupting service to equipment and circuits unrelated to those protected by the overcurrent device. Overcurrent devices are to be readily accessible. Overcurrent devices are not to be located where they could create an employee safety hazard by being exposed to physical damage, or located in the vicinity of easily ignitable material.

Fuses and Circuit Breakers

Fuses and circuit breakers must be so located or shielded that employees will not be burned or otherwise injured by their operation. Circuit breakers shall clearly indicate whether they are in the open (off) or closed (on) position. Where circuit breaker handles on switchboards are operated vertically rather than horizontally or rotationally, the up position of the handle is the closed (on) position. If used as switches in 120-volt, fluorescent lighting circuits and circuit breakers are to be marked "SWD." Feeders and branch circuits over 600 volts, nominal, must have short-circuit protection.

Grounding

The following systems which supply premises wiring are to be grounded such that all 3-wire DC systems have their neutral conductor grounded and two-wire DC systems, operating at over 50 volts through 300 volts between conductors, are to be grounded unless they are rectifier-derived from an AC system. AC circuits of less than 50 volts must be grounded if they are installed as overhead conductors outside of buildings, or if they are supplied by transformers and the transformer's primary supply system is ungrounded or exceeds 150 volts to ground. AC systems of 50 volts to 1000 volts are to be grounded under any of the following conditions, unless exempted:

1. If the system can be so grounded that the maximum voltage to ground on the ungrounded conductors does not exceed 150 volts.

2. If the system is nominally rated 480Y/277 volt, 3-phase, 4-wire in which the neutral is used as a circuit conductor.

3. If the system is nominally rated 240/120 volt, 3-phase, 4-wire in which the midpoint of one phase is used as a circuit conductor.

4. If a service conductor is uninsulated.

At certain times AC systems of 50 volts to 1000 volts are not required to be grounded, if the system is separately derived and is supplied by a transformer that has a primary voltage rating less than 1000 volts and provided all of the following conditions are met: the system is used exclusively to control circuits; the conditions of maintenance and supervision assure that only qualified persons will service the installation; the continuity of control power is required; and ground detectors are installed on the control system.

Separately Derived Systems

The requirements found in Ground Connections apply when the need exists for the grounding of wiring systems where power is derived from a generator, transformer, or converter windings, and there is no direct electrical connection, including a solidly connected grounded circuit conductor, to supply conductors originating in another system.

Portable- and Vehicle-Mounted Generators

Under the certain conditions, the frame of a portable generator need not be grounded and may serve as the grounding electrode for a system supplied by the generator (see Figure 29). Under these conditions the generator is to supply only equipment mounted on the generator and/or cord- and plug-connected equipment through receptacles mounted on the generator, and the noncurrent-carrying metal parts of equipment, and the equipment grounding conductor terminals of the receptacles, are to be bonded to the generator frame.

Figure 29. Example of portable electrical generator

For vehicle mounted generators, the frame of a vehicle may serve as the grounding electrode for a system supplied by a generator located on the vehicle, if certain conditions are met.

1. The frame of the generator is bonded to the vehicle frame.

2. The generator supplies only equipment located on the vehicle and/or cord- and plug-connected equipment through receptacles mounted on the vehicle or generator.

3. The noncurrent-carrying metal parts of equipment, and the equipment grounding conductor terminals of the receptacles, are bonded to the generator frame.

4. The system complies with all other provisions.

Neutral Conductor Bonding

A neutral conductor is to be bonded to the generator frame, if the generator is a component of a separately derived system. No other conductor need be bonded to the generator frame. For AC premises wiring systems the identified conductor must be grounded.

Ground Connections

For a grounded system, a grounding electrode conductor is used to connect both the equipment grounding conductor and the grounded circuit conductor to the grounding electrode. Both the equipment grounding conductor and the grounding electrode conductor is connected to the grounded circuit conductor on the supply side of the service disconnecting means, or on the supply side of the system disconnecting means or overcurrent devices, if the system is separately derived.

For an ungrounded service-supplied system, the equipment grounding conductor is connected to the grounding electrode conductor at the service equipment. For an ungrounded, separately derived system, the equipment grounding conductor is connected to the grounding electrode conductor at, or ahead of, the system disconnecting means or overcurrent devices. The path to ground from circuits, equipment, and enclosures is to be permanent and continuous.

Supports and Enclosures for Conductors

Metal cable trays, metal raceways, and metal enclosures for conductors are grounded, except that metal enclosures, such as sleeves, that are used to protect cable assemblies from physical damage need not be grounded. Metal enclosures for conductors, added to existing installations of open wire, knob-and-tube wiring, and nonmetallic-sheathed cable, need not be grounded, if all of the following conditions are met:

1.   Runs are less than 25 feet.

2.   Enclosures are free from probable contact with ground, grounded metal, metal laths, or other conductive materials.

3.   Enclosures are guarded against employee contact.

Metal enclosures for service equipment shall be grounded. Exposed noncurrent-carrying metal parts of fixed equipment, which may become energized, are to be grounded under any of the following conditions: if within 8 feet vertically or 5 feet horizontally of ground or grounded metal objects and subject to employee contact; if located in a wet or damp location and subject to employee contact; if in electrical contact with metal; if in a hazardous (classified) location; if supplied by a metal-clad, metal-sheathed, or grounded metal raceway wiring method; and if equipment operates with any terminal over 150 volts to ground. However, there is no need for grounding: if enclosures for switches or circuit breakers, are used for other than service equipment and are accessible to qualified persons only; if metal frames of electrically heated appliances are permanently and effectively insulated from ground; or if the cases of distribution apparatus, such as transformers and capacitors, are mounted on wooden poles at a height exceeding 8 feet aboveground or grade level.

Any exposed noncurrent-carrying metal parts of cord- and plug-connected equipment, which may become energized, are to be grounded: if located in a hazardous (classified) location; if operated at over 150 volts to ground, except for guarded motors and metal frames of electrically heated appliances, if the appliance frames are permanently and effectively insulated from ground; if the equipment is a hand-held motor-operated tool; if cord- and plug-connected equipment is used in damp or wet locations or is used by employees standing on the ground or metal floors or working inside metal tanks or boilers; if portable and mobile X-ray and associated equipment are used; if tools are likely to be used in wet and/or conductive locations; if portable hand lamps are used; or if tools are likely to be used in wet and/or conductive locations. They need not be grounded if supplied through an isolating transformer with an ungrounded secondary of not over 50 volts. Listed or labeled portable tools and appliances protected by a system of double insulation, or its equivalent, need not be grounded. If such a system is employed, the equipment shall be distinctively marked to indicate that the tool or appliance utilizes a system of double insulation.

Nonelectrical Equipment

The metal parts of the nonelectrical equipment are to be grounded. These include such things as frames and tracks of electrically operated cranes, frames of nonelectrically driven elevator cars to which electric conductors are attached, hand-operated metal shifting ropes or cables of electric elevators, and metal partitions, grill work, and similar metal enclosures around equipment of over 1kV between conductors.

Noncurrent-carrying metal parts of fixed equipment, if required to be grounded, are grounded by an equipment grounding conductor which is contained within the same raceway, cable, or cord, or runs with or encloses the circuit conductors. For DC circuits only, the equipment grounding conductor may be run separately from the circuit conductors. A conductor used for grounding fixed or movable equipment must have the capacity to conduct safely any fault current which may be imposed on it.

Effective Grounding

Electric equipment is considered to be effectively grounded if it is secured to, and in electrical contact with, a metal rack or structure that is provided for its support. The metal rack or structure is to be grounded by the method specified for the noncurrent carrying metal parts of fixed equipment, or metal car frames are to be supported by metal hoisting cables attached to, or running over metal sheaves or drums of grounded elevator machines.

Bonded Conductors

Bonding conductors are used to assure electrical continuity. They have the capacity to conduct any fault current which may be imposed.

Made Electrodes

If made electrodes are used, they are to be free from nonconductive coatings, such as paint or enamel, and, if practicable, they are to be embedded below the permanent moisture level. A single electrode, consisting of a rod, pipe, or plate, which has a resistance to ground greater than 25 ohms, must be augmented by one additional electrode and installed no closer than 6 feet to the first electrode.

Grounded High Voltage

The grounding of systems and circuits of 1000 volts and over (high voltage) are usually in compliance, if they meet the previous grounding requirements. When grounding systems that supply portable or mobile equipment, the systems which are supplying portable or mobile high voltage equipment, other than substations installed on a temporary basis, must assure that the portable and mobile high voltage equipment is supplied from a system having its neutral grounded through an impedance. If a delta-connected high voltage system is used to supply the equipment, a system neutral shall be derived. Also, exposed noncurrent-carrying metal parts of portable and mobile equipment are to be connected by an equipment grounding conductor to the point at which the system's neutral impedance is grounded.

The use of ground-fault detection and relaying shall be provided and it shall automatically deenergize any high voltage system component which has developed a ground fault. The continuity of the equipment grounding conductor must be continuously monitored so as to deenergize, automatically, the high voltage feeder to the portable equipment upon loss of continuity of the equipment grounding conductor. The grounding electrode, to which the portable or mobile equipment system neutral impedance is connected, is to be isolated from, and separated in the ground by at least 20 feet from any other system or equipment grounding electrode; there must be no direct connection between the grounding electrodes, such as buried pipe, fence, or like objects. All noncurrent-carrying metal parts of portable equipment and fixed equipment, including their associated fences, housings, enclosures, and supporting structures, are to be grounded.

However, equipment which is guarded by location, and isolated from ground, need not be grounded. Additionally, pole-mounted distribution apparatus, at a height exceeding 8 feet aboveground or grade level, need not be grounded.

Wiring Methods, Components, and Equipment for General Use (1926.405)

The requirements in this section do not apply to conductors which form an integral part of equipment such as motors, controllers, motor control centers, and like equipment.

General Requirements

Metal raceways, cable armor, and other metal enclosures for conductors are metallically joined together into a continuous electric conductor and are so connected to all boxes, fittings, and cabinets as to provide effective electrical continuity. No wiring systems of any type are to be installed in ducts used to transport dust, loose stock, or flammable vapors, and no wiring systems of any type are to be installed in any duct used for vapor removal, or in any shaft containing only such ducts.

Temporary Wiring

Temporary electrical power and lighting wiring methods which may be of a class less than would be required for a permanent installation. Except as noted, all other requirements for permanent wiring apply to temporary wiring installations. Temporary wiring is removed immediately upon completion of construction, or removed once the purpose for which the wiring was installed is completed.

General Requirements for Temporary Wiring

Feeders are to originate in a distribution center. The conductors must run as multiconductor cord or cable assemblies, or within raceways. Where they are not subject to physical damage, they may be run as open conductors on insulators that are not more than 10 feet apart.

Branch circuits are to originate in a power outlet or panelboard. Again, conductors run as multiconductor cord or cable assemblies, open conductors, or run in raceways. All conductors are to be protected by overcurrent devices at their ampacity. Runs of open conductors shall be located where the conductors will not be subject to physical damage, and the conductors are to be fastened at intervals not exceeding 10 feet. No branch-circuit conductors shall be laid on the floor. Each branch circuit that supplies receptacles or fixed equipment must contain a separate equipment grounding conductor, if the branch circuit is run as open conductors.

Receptacles Must Be of the Grounding Type

Unless installed in a complete metallic raceway, each branch circuit must contain a separate equipment grounding conductor, and all receptacles are to be electrically connected to the grounding conductor. Receptacles, used for other than temporary lighting, must not be installed on branch circuits which supply temporary lighting. Receptacles shall not be connected to the same ungrounded conductor of multiwire circuits which supply temporary lighting.

Disconnecting switches or plug connectors shall be installed to permit the disconnection of all ungrounded conductors of each temporary circuit.

Temporary Lights

All lamps used for general illumination are to be protected from accidental contact or breakage. Metal-case sockets must be grounded. Temporary lights shall not be suspended by their electric cords unless cords and lights are designed for this means of suspension (see Figure 30). Portable electric lighting used in wet and/or other conductive locations, as for example, drums, tanks, and vessels, are to be operated at 12 volts or less. However, 120-volt lights may be used if protected by a ground-fault circuit interrupter.

Figure 30. Protected temporary lighting

Boxes

A box is used wherever a change is made to a raceway system or cable system which is metal clad or metal sheathed.

Flexible Cords and Cables

Flexible cords and cables are to be protected from damage. Sharp corners and projections need to be avoided. Flexible cords and cables may pass through doorways or other pinch points, if protection is provided to avoid damage. Extension cord sets, used with portable electric tools and appliances, must be of three-wire type and designed for hard or extra-hard usage. Flexible cords, used with temporary and portable lights, shall be designed for hard or extra-hard usage. The types of flexible cords designed for hard or extra-hard usage are: hard service types S, ST, SO, STO, and junior hard service cord types SJ, SJO, SJT, SJTO; these are acceptable for use on construction sites.

Flexible cords and cables are to be suitable for the conditions of use and location. Flexible cords and cables are only used for pendants, wiring of fixtures, connection of portable lamps or appliances, elevator cables, wiring of cranes and hoists, connection of stationary equipment to facilitate their frequent interchange, prevention of the transmission of noise or vibration, and appliances where the fastening means and mechanical connections are designed to permit removal for maintenance and repair. The attachment plugs for flexible cords, if used, are to be equipped with an attachment plug and energized from a receptacle outlet.

Flexible cords and cables are not to be used as a substitute for the fixed wiring of a structure, for runs through holes in walls, ceilings, or floors, for a run through doorways, windows, or similar openings, for attachment to building surfaces, or for concealment behind building walls, ceilings, or floors. A conductor of a flexible cord or cable, that is used as a grounded conductor or an equipment grounding conductor, is to be distinguishable from other conductors.

Flexible cords are used only in continuous lengths without splice or tap. Hard service flexible cords, No. 12 or larger, may be repaired if spliced so that the splice retains the insulation, outer sheath properties, and usage characteristics of the cord being spliced. Flexible cords are to be connected to devices and fittings so that strain relief is provided which prevents the pull from being directly transmitted to joints or terminal screws. Flexible cords and cables must be protected by bushings or fittings when they are passed through holes in covers, outlet boxes, or similar enclosures.

Guarding

For temporary wiring over 600 volts, nominal, there must be fencing, barriers, or other effective means provided to prevent access of other than authorized and qualified personnel (see Figure 31).

Cabinets, Boxes, and Fittings

Conductors entering boxes, cabinets, or fittings are to be protected from abrasion, and openings through which conductors enter must be effectively closed. Unused openings in cabinets, boxes, and fittings must also be effectively closed.

All pull boxes, junction boxes, and fittings shall be provided with covers. If metal covers are used, they are grounded. In energized installations, each outlet box shall have a cover, faceplate, or fixture canopy. Covers of outlet boxes which have holes through which flexible cord pendants passed, must provide bushings designed for the purpose, or have smooth, well-rounded surfaces on which the cords may bear.

In addition to other requirements for pull and junction boxes, the following applies for systems over 600 volts, nominal:

1. Boxes are to be provided a complete enclosure for the contained conductors or cables.

2. Boxes shall be closed by covers securely fastened in place. Underground box covers that weigh over 100 pounds must meet this requirement. Covers for boxes must be permanently marked "HIGH VOLTAGE." The marking is to be on the outside of the box cover and be readily visible and legible.

Figure 31. Typical dog house as a barrier from electricity for unqualified persons

Knife Switches

Single-throw knife switches are to be connected so that the blades are dead when the switch is in the open position. Single-throw knife switches are to be placed so that gravity will not tend to close them. Single-throw knife switches, approved for use in the inverted position, are to provide a locking device that will ensure that the blades remain in the open position, when so set. Double-throw knife switches may be mounted so that the throw will be either vertical or horizontal. However, if the throw is vertical, a locking device must be provided to ensure that the blades remain in the open position, when so set. Exposed blades of knife switches must be dead when open.

Switchboards and Panelboards

Switchboards that have any exposed live parts are to be located in permanently dry locations and accessible only to qualified persons. Panelboards shall be mounted in cabinets, cutout boxes, or enclosures designed for the purpose and must be dead front. However, panelboards, other than the dead front externally-operable type, are permitted where accessible only to qualified persons.

Wet or Damp Locations

Cabinets, cutout boxes, fittings, boxes, and panelboard enclosures, which are in damp or wet locations, must be installed so as to prevent moisture or water from entering and accumulating within the enclosures. In wet locations, the enclosures are to be weatherproof. Switches, circuit breakers, and switchboards, installed in wet locations, must be enclosed in weatherproof enclosures.

Conductors

All conductors used for general wiring are to be insulated with few exceptions. The conductor insulation is to be of a type that is suitable for the voltage, operating temperature, and location of use. Insulated conductors must be distinguished as being grounded conductors, ungrounded conductors, or equipment grounding conductors by using appropriate colors, or other means. Multiconductor portable cable, used to supply power to portable or mobile equipment at over 600 volts, nominal, must consist of No. 8, or larger, conductors employing flexible stranding. Cables operated at over 2000 volts are to be shielded for the purpose of confining the voltage stresses to the insulation, and grounding conductors must be provided. Connectors for these cables are to be of a locking type, with provisions to prevent their opening or closing while energized. Strain relief must be provided at connections and terminations. Portable cables are not operated with splices unless the splices are of the permanent molded, vulcanized, or other equivalent type. Termination enclosures must be marked with a high voltage hazard warning, and terminations shall be accessible only to authorized and qualified personnel.

Fixtures

Fixture wires must be suitable for the voltage, temperature, and location of use. A fixture wire which is used as a grounded conductor is to be identified. Fixture wires may be used for installation in lighting, fixtures, and similar equipment when enclosed or protected, when not subject to bending or twisting while in use, and when used for connecting lighting fixtures to the branch-circuit conductors supplying the fixtures. Fixture wires are not designed as branch-circuit conductors except as permitted for Class 1 power-limited circuits.

Fixtures, lampholders, lamps, rosettes, and receptacles are, normally, to have no live parts exposed to employee contact. However, rosettes and cleat-type lampholders, and receptacles located at least 8 feet above the floor, may have exposed parts. Fixtures, lampholders, rosettes, and receptacles are to be securely supported. A fixture that weighs more than 6 pounds, or exceeds 16 inches in any dimension, shall not be supported by the screw shell of a lampholder.

Portable lamps are to be wired with flexible cord and have an attachment plug of the polarized or grounding type. If the portable lamp uses an Edison-based lampholder, the grounded conductor shall be identified and attached to the screw shell and the identified blade of the attachment plug. In addition, portable handlamps must have a metal shell; paperlined lampholders are not to be used; handlamps must be equipped with a handle of molded composition or other insulating material; handlamps are to be equipped with a substantial guard attached to the lampholder or handle; and metallic guards shall be grounded by the means of an equipment grounding conductor which runs within the power supply cord.

Lampholders of the screw-shell type are to be installed for use as lampholders only. Lampholders installed in wet or damp locations must be of the weatherproof type. Fixtures installed in wet or damp locations must be identified for the purpose and installed so that water cannot enter or accumulate in wireways, lampholders, or other electrical parts.

Receptacles

Receptacles, cord connectors, and attachment plugs are to be constructed so that no receptacle or cord connector will accept an attachment plug with a different voltage or currentrating, other than that for which the device is intended. However, a 20-ampere T-slot receptacle or cord connector may accept a 15-ampere attachment plug of the same voltage rating. Receptacles connected to circuits, having different voltages, frequencies, or types of current (ac or dc) on the same premises, must be of such design that the attachment plugs used on these circuits are not interchangeable. A receptacle installed in a wet or damp location must be designed for the location.

Appliances

Appliances, other than those in which the current-carrying parts at high temperatures are necessarily exposed, shall, normally, have no live parts exposed to employee contact. A means is to be provided to disconnect each appliance. Each appliance is to be marked with its rating in volts and amperes, or volts and watts.

Motors

Motors, motor circuits, and controllers may require that one piece of equipment shall be "in sight from" another piece of equipment: one is visible and not more than 50 feet from the other. A disconnecting means is to be located, in sight, from the controller location. The controller disconnecting means for motor branch circuits over 600 volts, nominal, may be out of sight of the controller, if the controller is marked with a warning label that gives the location and identification of the disconnecting means; this means is to be locked in the open position.

The disconnecting means must disconnect the motor and the controller from all ungrounded supply conductors, and be designed so that no pole can be operated independently. If a motor and the driven machinery are not in sight from the controller location, the installation must comply with one of the following conditions: the controller disconnecting means must be capable of being locked in the open position; a manually operable switch must be placed in sight of the motor location that will disconnect the motor from its source of supply; the disconnecting means must plainly indicate whether it is in the open (off) or closed (on) position; the disconnecting means must be readily accessible; if more than one disconnect is provided for the same equipment, only one need be readily accessible; or an individual disconnecting means must be provided for each motor.

For a specific group of motors, a single disconnecting means may be used under any one of the following conditions:

1. If a number of motors drive special parts of a single machine or piece of apparatus, such as a metal or woodworking machine, crane, or hoist.

2. If a group of motors is under the protection of one set of branch-circuit protective devices.

3. If a group of motors is in a single room in sight from the location of the disconnecting means.

Motors, motor-control apparatus, and motor branch-circuit conductors are to be protected against overheating due to motor overloads, failure to start, short circuits, or ground faults. These provisions do not require overload protection that will stop a motor, when a shutdown is likely to introduce additional or increased hazards, as in the case of fire pumps, or where continued operation of a motor is necessary for a safe shutdown of equipment, or where process and motor overload sensing devices are connected to a supervised alarm.

Stationary motors, which have commutators, collectors, and brush rigging located inside of motor end brackets and are not conductively connected to supply circuits operating at more than 150 volts to ground, need not have such parts guarded. Exposed live parts of motors and controllers, operating at 50 volts or more between terminals, are to be guarded against accidental contact by any of the following:

1. By installation in a room or enclosure that is accessible only to qualified persons.

2. By installation on a balcony, gallery, or platform, so elevated and arranged as to exclude unqualified persons.

3. By elevation 8 feet or more above the floor.

Where there are live parts of motors, or controllers are operating at over 150 volts to ground, they are to be guarded against accidental contact only by location. And, where adjustments or other attendance may be necessary during the operation of the apparatus, insulating mats or platforms are to be provided so that the attendant cannot readily touch live parts, unless standing on the mats or platforms.

Transformers

The only transformers that are not addressed in this section are the current transformers, dry-type transformers which are installed as a component part of other apparatus, transformers which are an integral part of an X-ray, high frequency, or electrostatic-coating apparatus, or transformers used with Class 2 and Class 3 circuits, signs, outline lighting, electric discharge lighting, and power-limited fire-protective signaling.

The operating voltage of exposed live parts of transformer installations are to be indicated by warning signs or visible markings on the equipment or structure. Dry-type, high fire point liquid-insulated, and askarel-insulated transformers installed indoors and rated over 35 kV must be placed in a vault. If oil-insulated transformers present a fire hazard to employees, they are to be installed indoors in a vault.

Fire Protection

Combustible material, combustible buildings and parts of buildings, fire escapes, and door and window openings are to be safeguarded from fires which may originate in oil-insulated transformers attached to, or adjacent to a building, or combustible material. Transformer vaults are to be constructed so as to contain fire and combustible liquids within the vault and to prevent unauthorized access. Locks and latches must be so arranged that a vault door can be readily opened from the inside.

Transformer Guidelines

Any pipe or duct system which is foreign to the vault installation shall not enter or pass through a transformer vault. Materials must not be stored in transformer vaults.

Capacitors

All capacitors, except surge capacitors or capacitors included as a component part of other apparatus, are to be provided with an automatic means of draining the stored charge and maintaining the discharged state after the capacitor is disconnected from its source of supply. Capacitors rated over 600 volts, nominal, have other requirements and, therefore, need isolating or disconnecting switches (with no interrupting rating) which are interlocked with a load interrupting device or provided with prominently displayed caution signs that prevent switching load current. For series capacitors, proper switching must be assured by using at least mechanically sequenced isolating and bypass switches, interlocks, or a switching procedure which is prominently displayed at the switching location.

Specific Purpose Equipment and Installation (1926.406)

The installation of electric equipment and wiring used in connection with cranes, monorail hoists, hoists, and all runways is addressed in this section. A readily accessible disconnecting means is to be provided between the runway contact conductors and the power supply. The disconnecting means must be capable of being locked in the open position, and is provided in the leads from the runway contact conductors or other power supply on any crane or monorail hoist. If this additional disconnecting means is not readily accessible from the crane or monorail hoist operating station, a means must be provided at the operating station to open the power circuit to all motors of the crane or monorail hoist. The additional disconnect may be omitted if a monorail hoist or hand-propelled crane bridge installation meets all of the following:

1. The unit is floor controlled.

2. The unit is within view of the power supply disconnecting means.

3. No fixed work platform has been provided for servicing the unit.

A limit switch or other device shall be provided to prevent the load block from passing the safe upper limit of travel of any hoisting mechanism. The dimension of the working space in the direction of access to live parts which may require examination, adjustment, servicing, or maintenance while alive, is to be a minimum of 2 feet 6 inches. Where controls are enclosed in cabinets, the door(s) are to open at least 90 degrees, be removable, or the installation must provide equivalent access.

All exposed metal parts of cranes, monorail hoists, hoists, and accessories, including pendant controls, are to be metallically joined together into a continuous electrical conductor so that the entire crane or hoist will be grounded. Moving parts, other than removable accessories or attachments, having metal-to-metal bearing surfaces are considered to be electrically connected to each other through the bearing surfaces for grounding purposes. The trolley frame and bridge frame are considered as electrically grounded through the bridge and trolley wheels and its respective tracks unless conditions, such as paint or other insulating materials, prevent reliable metal-to-metal contact. In this case, a separate bonding conductor shall be provided.

Elevators, Escalators, and Moving Walks

Elevators, escalators, and moving walks must have a single means for disconnecting all ungrounded main power supply conductors for each unit. If control panels are not located in the same space as the drive machine, they are to be located in cabinets with doors or panels capable of being locked closed.

Electric Welder Disconnects

Motor-generators, AC transformers, and DC rectifier arc welders must have a disconnecting means, in the supply circuit, for each motor-generator arc welder, and for each AC transformer and DC rectifier arc welder, which is not equipped with a disconnect that is mounted as an integral part of the welder. A switch or circuit breaker is to be provided by which each resistance welder and its control equipment can be isolated from the supply circuit. The ampere rating of this disconnecting means must not be less than the supply conductor ampacity.

X-ray Equipment

A disconnecting means for X-ray equipment is to be provided in the supply circuit. The disconnecting means is to be operable from a location readily accessible from the X-ray control. For equipment connected to a 120-volt branch circuit of 30 amperes or less, a grounding-type attachment plug cap and receptacle of proper rating may serve as a disconnecting means. If more than one piece of equipment is operated from the same high-voltage circuit, each piece or each group of equipment as a unit must be provided with a high-voltage switch or equivalent disconnecting means. This disconnecting means shall be constructed, enclosed, or located so as to avoid contact by employees with its live parts. All radiographic and fluoroscopic-type equipment are to be effectively enclosed or have interlocks that de-energize the equipment automatically to prevent ready access to live current-carrying parts.

Hazardous (Classified) Locations (1926.407)

The requirements for electric equipment and wiring depends on the location of the properties and the potential for the presence of a flammable or combustible concentration of vapors, liquids, gases, combustible dusts, or fibers. Each room, section, or area shall be considered individually in determining its classification. These hazardous (classified) locations are assigned six designations (see Figure 32).

Class I, Division 1
Class I, Division 2
Class II, Division 1
Class II, Division 2
Class III, Division 1
Class III, Division 2

(See Appendix H for detailed description on hazard locations.)

Equipment, wiring methods, and installations of equipment in hazardous (classified) locations are to be approved as intrinsically safe, or approved for the hazardous (classified) location. Requirements for each of these options are that the equipment and associated wiring, which is approved as intrinsically safe, is permitted in any hazardous (classified) location included in its listing or labeling, and is approved for the hazardous (classified) location.

Equipment must be approved not only for the class of location, but also for the ignitable or combustible properties of the specific gas, vapor, dust, or fiber that will be present. The NFPA 70, the National Electrical Code, lists or defines hazardous gases, vapors, and dusts by "Groups" characterized by their ignitable or combustible properties.

Equipment must not be used unless it is marked to show the class, group, and operating temperature or temperature range, based on operation in a 40-degree C ambient, for which it is approved. The temperature marking must not exceed the ignition temperature of the specific gas, vapor, or dust to be encountered. However, the following provisions modify this marking requirement for specific equipment:

1. Equipment of the non-heat-producing type (such as junction boxes, conduit, and fittings) and equipment of the heat-producing type having a maximum temperature of not more than 100 degrees C (212 degrees F) need not have a marked operating temperature or temperature range.

2. Fixed lighting fixtures marked for use only in Class I, Division 2 locations need not be marked to indicate the group.

3. Fixed general-purpose equipment in Class I locations, other than lighting fixtures, which is acceptable for use in Class I, Division 2 locations need not be marked with the class, group, division, or operating temperature.

4. Fixed dust-tight equipment, other than lighting fixtures, which is acceptable for use in Class II, Division 2 and Class III locations need not be marked with the class, group, division, or operating temperature.

Equipment which is safe for the location is to be of a type and design which the employer demonstrates will provide protection from the hazards arising from the combustibil-

Summary of Class I, II, III Hazardous Locations			
CLASSES	**GROUPS**	**DIVISIONS**	
		1	**2**
I Gases, vapors, and liquids (Art. 501)	A: Acetylene B: Hydrogen, etc. C: Ether, etc. D: Hydrocarbons, fuels, solvents, etc.	Normally explosive and hazardous	Not normally present in an explosive concentration (but may accidentally exist)
II Dusts (Art. 502)	E: Metal dusts (conductive,* and explosive) F: Carbon dusts (some are conductive, and all are explosive) G: Flour, starch, grain, combustible plastic or chemical dust (explosive)	Ignitable quantities of dust normally are or may be in suspension, or conductive dust may be present	Dust not normally suspended in an ignitable concentration (but may accidentally exist). Dust layers are present.
III Fibers and flyings (Art. 503)	Textiles, wood-working, etc. (easily ignitable, but not likely to be explosive)	Handled or used in manufacturing	Stored or handled in storage (exclusive of manufacturing)

Figure 32. Hazardous locations

ity and flammability of vapors, liquids, gases, dusts, or fibers. The National Electrical Code, NFPA 70, contains a guidelines form determining the type and design of equipment and installations which will meet this requirement.

All conduits are to be threaded and made wrench-tight. Where it is impractical to make a threaded joint tight, a bonding jumper shall be utilized.

Special Systems (1926.408)

Systems over 600 volts, nominal, must meet the general requirements for all circuits and equipment operated at over 600 volts. Wiring methods for fixed installations containing above-ground conductors are to be installed in rigid metal conduit, in intermediate metal conduit, in cable trays, in cable bus, in other suitable raceways, or as open runs of metal-clad cable designed for this use and purpose. However, open runs of non-metallic-sheathed cable, or of bare conductors or busbars, may be installed in locations which are accessible only to qualified persons. Metallic shielding components, such as tapes, wires, or braids for conductors, are to be grounded. Open runs of insulated wires and cables having a bare lead sheath or a braided outer covering must be supported in a manner designed to prevent physical damage to the braid or sheath.

Installations Emerging from the Ground

Conductors emerging from the ground need to be enclosed in raceways. Raceways installed on poles are to be of rigid metal conduit, intermediate metal conduit, PVC schedule 80, or the equivalent and extend from the ground line up to a point 8 feet above finished grade. Conductors entering a building shall be protected by an enclosure from the ground line to the point of entrance. Metallic enclosures are to be grounded.

Interrupting and Isolating Devices

Circuit breakers located indoors must consist of metal-enclosed or fire-resistant, cell-mounted units. In locations accessible only to qualified personnel, open mounting of circuit breakers is permitted. A means of indicating the open and closed position of circuit breakers is to be provided.

Fused cutouts, installed in buildings or transformer vaults, are to be of a type identified for the purpose. They shall be readily accessible for fuse replacement. A means must be provided to completely isolate equipment for inspection and repairs. Isolating means which are not designed to interrupt the load current of the circuit shall be either interlocked with a circuit interrupter or provided with a sign warning against opening them under load.

Mobile and Portable Equipment

A metallic enclosure is to be provided on the mobile machine for enclosing the terminals of the power cable. The enclosure must include provisions for a solid connection for the ground wire(s) terminal to ground, effectively, the machine frame. The method used for cable termination must prevent any strain or pull on the cable from stressing the electrical connections. The enclosure is to have provisions for locking so that only authorized qualified persons may open it, and it must be marked with a sign warning of the presence of energized parts.

Guarding Live Parts

All energized switching and control parts are to be enclosed in effectively grounded metal cabinets or enclosures. Circuit breakers and protective equipment shall have the operating means projecting through the metal cabinet or enclosure so these units can be reset without the locked doors being opened. Enclosures and metal cabinets are to be locked so that only authorized qualified persons have access, and must be marked with a sign warning of the presence of energized parts. Collector ring assemblies on revolving-type machines (shovels, draglines, etc.) are to be guarded.

Tunnel Installations

The installation and use of high-voltage power distribution and utilization equipment which are associated with tunnels and are portable and/or mobile, such as substations, trailers, cars, mobile shovels, draglines, hoists, drills, dredges, compressors, pumps, conveyors, and underground excavators, are to have conductors installed in one or more of the following:

1. Metal conduit or other metal raceway.

2. Type MC cable.

3. Other suitable multiconductor cable. Conductors are also to be so located or guarded as to protect them from physical damage. Multiconductor portable cables may supply mobile equipment. An equipment grounding conductor must run with circuit conductors inside the metal raceway or inside the multiconductor cable jacket. The equipment grounding conductor may be insulated or bare.

Bare terminals of transformers, switches, motor controllers, and other equipment are to be enclosed to prevent accidental contact with energized parts. Enclosures for use in tunnels are to be drip-proof, weatherproof, or submersible as required by the environmental conditions.

A disconnecting means that simultaneously opens all ungrounded conductors is to be installed at each transformer or motor location. All nonenergized metal parts of electric equipment and metal raceways and cable sheaths shall be grounded and bonded to all metal pipes and rails at the portal and at intervals not exceeding 1000 feet throughout the tunnel.

Classification. Class 1, Class 2, or Class 3 Remote Control, Signaling, or Power-Limited Circuits

Class 1, Class 2, or Class 3 remote control, signaling, or power-limited circuits are characterized by their usage and electrical power limitation which differentiates them from light and power circuits. These circuits are classified in accordance with their respective voltage and power limitations.

A Class 1 power-limited circuit is to be supplied from a source having a rated output of not more than 30 volts and 1000 volt-amperes. A Class 1 remote control circuit or a Class 1 signaling circuit shall have a voltage which does not exceed 600 volts; however, the power output of the source need not be limited.

Power for Class 2 and Class 3 circuits shall be limited either inherently (in which no overcurrent protection is required), or by a combination of a power source and overcurrent protection. The maximum circuit voltage is to be 150 volts AC or DC for a Class 2 inherently limited power source, and 100 volts AC or DC for a Class 3 inherently limited power source. The maximum circuit voltage is to be 30 volts AC and 60 volts DC for a Class 2 power source limited by overcurrent protection, and 150 volts AC or DC for a Class 3 power source limited by overcurrent protection. The maximum circuit voltages apply to sinusoidal AC or continuous DC power sources, and where wet contact occurrence is not likely. A Class 2 or Class 3 power supply unit is not to be used unless it is durably marked and plainly visible to indicate the class of supply and its electrical rating.

Communications Systems

The provisions for communication systems apply to such systems as central-station-connected and non-central-station-connected telephone circuits, radio receiving and transmitting equipment, outside wiring for fire and burglar alarm, and similar central station systems. Communication circuits so located as to be exposed to accidental contact with light or power conductors operating at over 300 volts, must have each circuit, so exposed, provided with an approved protector.

Each conductor of a lead-in from an outdoor antenna must be provided with an antenna discharge unit or other means that will drain static charges from the antenna system. Receiving distribution lead-in or aerial-drop cables attached to buildings, and lead-in conductors to radio transmitters, are to be so installed as to avoid the possibility of accidental contact with electric light or power conductors. The clearance between lead-in conductors and any lightning protection conductors is to be not less than 6 feet. Where practicable, communication conductors on poles shall be located below the light or power conductors. Communications conductors must not be attached to a crossarm that carries light or power conductors.

Indoor antennas, lead-ins, and other communication conductors attached as open conductors to the inside of buildings are to be located at least 2 inches from conductors of any light, power, or Class 1 circuits unless a special and equally protective method of conductor separation is employed.

Outdoor metal structures which support antennas, as well as self-supporting antennas such as vertical rods or dipole structures, are to be located as far away from overhead conductors of electric light and power circuits of over 150 volts to ground, as necessary, to avoid the possibility of the antenna or structure falling into, or making accidental contact with such circuits.

If lead-in conductors are exposed to contact with electric light or power conductors, the metal sheath of aerial cables entering buildings must be grounded or interrupted close to the entrance to the building by an insulating joint or equivalent device. Where protective devices are used, they are to be grounded. Masts and metal structures supporting antennas must be permanently and effectively grounded without splice or connection in the grounding conductor.

Transmitters are to be enclosed in a metal frame or grill, or separated from the operating space by a barrier, and all metallic parts shall be effectively connected to ground. All external metal handles and controls which are accessible to the operating personnel shall be effectively grounded. Unpowered equipment and enclosures are considered grounded when connected to an attached coaxial cable with an effectively grounded metallic shield.

Electrical Work Practices (1926.416)

No employer shall permit an employee to work in proximity of any part of an electric power circuit where the employee could contact the electric power circuit in the course of work, unless the employee is protected against electric shock by deenergizing the circuit and grounding it, or by guarding it effectively by insulation or other means.

In work areas where the exact location of underground electric powerlines is unknown, employees using jack-hammers, bars, or other hand tools which may contact a line, are to be provided with insulated protective gloves.

Before work is begun, the employer must ascertain, by inquiry, direct observation, or instruments, whether any part of an energized electric power circuit, exposed or concealed, is so located that the performance of the work may bring any person, tool, or machine into physical or electrical contact with the electric power circuit. The employer must post and maintain proper warning signs where such a circuit exists. The employer shall advise employees of the location of such lines, the hazards involved, and the protective measures to be taken.

Barriers, or other means of guarding are to be provided to ensure that the workspace for electrical equipment will not be used as a passageway during periods when energized parts of electrical equipment are exposed. Working spaces, walkways, and similar locations are to be kept clear of cords so as not to create a hazard to employees.

In existing installations, no changes in circuit protection shall be made to increase the load in excess of the load rating of the circuit wiring. When fuses are installed or removed with one or both terminals energized, special tools insulated for the voltage are to be used.

Worn or frayed electric cords or cables must not be used. Extension cords shall not be fastened with staples, hung from nails, or suspended by wire.

Lockout/Tagging of Circuits (1926.417)

Controls that are to be deactivated during the course of work on energized or deenergized equipment or circuits are to be tagged. Equipment or circuits that are deenergized are to be rendered inoperative and have tags attached at all points where such equipment or circuits can be energized. Tags must be placed to identify plainly the equipment or circuits being worked on (see Figure 33).

Figure 33. Lockout/tagout example

Safety-Related Maintenance and Environmental Considerations

Maintenance of Equipment (1926.431)

The employer shall ensure that all wiring components and utilization equipment in hazardous locations are maintained in a dust-tight, dust-ignition-proof, or explosion-proof condition, as appropriate. There shall be no loose or missing screws, gaskets, threaded connections, seals, or other impairments to a tight condition.

Environmental Deterioration of Equipment (1926.432)

Unless identified for use in the operating environment, no conductors or equipment shall be located

1. Where locations are damp or wet.

2. Where exposed to gases, fumes, vapors, liquids, or other agents having a deteriorating effect on the conductors or equipment.

3. Where exposed to excessive temperatures.

Control equipment, utilization equipment, and busways, approved for use in dry locations only, are protected against damage from the weather during building construction.

Metal raceways, cable armor, boxes, cable sheathing, cabinets, elbows, couplings, fittings, supports, and support hardware are of materials appropriate for the environment in which they are to be installed.

EMPLOYEE EMERGENCY ACTION PLANS (1926.35)

The emergency action plan is to be in writing and cover those designated actions employers and employees must take to ensure employee safety from fire and other emergencies. The emergency action plan must contain these elements, at a minimum:

1. Emergency escape procedures and emergency escape route assignments.

2. Procedures to be followed by employees who remain to operate critical plant operations before they evacuate.

3. Procedures to account for all employees after emergency evacuation has been completed.

4. Rescue and medical duties for those employees who are to perform them.

5. The preferred means of reporting fires and other emergencies.

6. Names or regular job titles of persons or departments who can be contacted for further information or explanation of duties under the plan.

The employer must establish an employee alarm system which complies with 1926.159. If the employee alarm system is used for alerting fire brigade members, or for other purposes, a distinctive signal for each purpose is to be used.

Before implementing the emergency action plan, the employer must designate and train a sufficient number of persons to assist in the safe and orderly emergency evacuation of employees. The employer must review the plan with each employee covered by the plan at the following times:

1. Initially when the plan is developed.

2. Whenever the employee's responsibilities or designated actions under the plan change.

3. Whenever the plan is changed.

The employer must review with each employee, upon initial assignment, those parts of the plan which the employee must know to protect the employee in the event of an emergency. The written plan is to be kept at the workplace and made available for employee review. For those employers with 10 or fewer employees, the plan may be communicated orally to employees and the employer need not maintain a written plan.

EXCAVATIONS /TRENCHES (1926.650)

Excavation operations are among the first undertaken at a construction site. Accidental cave-ins of the earth that has been excavated account for a large number of construction site related deaths. All openings made in the earth's surface are considered excavations. Excavations are defined to include trenches.

In almost all cases of excavation/trench accidents, the resulting accident occurred because the known regulations and safe work practices were violated. This is not a statement which can be made about most construction accidents.Trenching and excavation work presents a serious risk to all employees. The greatest risk, and one of primary concern, is a cave-in. A cubic yard of soil, or other material weighs more than 2000 pounds and usually more than a cubic yard of material is involved. When a cave-in occurs, this type of weight is beyond the physical capabilities of workers to protect themselves. Cave-in accidents are much more likely to result in worker fatalities than any other excavation-related accident (see Figure 34).

Specific Excavation Requirements (1926.651)

All surface encumbrances, that are located so as to create a hazard to employees, are to be removed or supported, as necessary, to safeguard employees.

Figure 34. Workers in an unprotected trench

Utilities

The estimated location of utility installations, such as the sewer, telephone, fuel, electric, or water lines, or any other underground installation that may be expected to be encountered during excavation work, is to be determined prior to opening an excavation. Prior to the start of the actual excavation, utility companies or owners must be contacted within the established or customary local response times, advised of the proposed work, and asked to establish the location of the utility underground installations. When utility companies or owners cannot respond to a request to locate underground utility installations within 24 hours (unless a longer period is required by state or local law), or cannot establish the exact location of these installations, the employer may proceed, provided the employer does so with caution, and provided detection equipment, or other acceptable means to locate utility installations are used. When excavation operations approach the estimated location of underground installations, the exact location of the installations shall be determined by safe and acceptable means. While the excavation is open, underground installations are to be protected, supported, or removed, as necessary, to safeguard employees (see Figure 35).

Egress Ramps and Runways

Structural ramps, that are used solely by employees as a means of access or egress from excavations, are to be designed by a competent person. Structural ramps, used for access or egress of equipment, are to be designed by a competent person qualified in structural design, and are to be constructed in accordance with the design. Ramps and runways constructed of two or more structural members must have the structural members connected together to prevent displacement. Structural members, used for ramps and runways, must be of uniform thickness. Cleats, or other appropriate means used to connect runway structural members, are to be attached to the bottom of the runway, or attached in a manner to prevent tripping. Structural ramps, used in lieu of steps, must provide cleats or other surface treatments on the top surface to prevent slipping. A stairway, ladder, ramp, or other safe means of egress must be located in trench excavations that are 4 feet or more in depth so as to require no more than 25 feet of lateral travel for employees (see Figure 36).

Figure 35. Unsupported trench and underground installations

Equipment and Loads

     No worker is permitted underneath loads handled by lifting or digging equipment. To avoid being struck by any spillage or falling materials, employees are required to stand away from any vehicle being loaded or unloaded. Operators may remain in the cabs of vehicles being loaded or unloaded when the vehicles are equipped to provide adequate protection for the operator during loading and unloading operations.

     When mobile equipment is operated adjacent to an excavation, or when such equipment is required to approach the edge of an excavation, and the operator does not have a clear and direct view of the edge of the excavation, a warning system is to be utilized; these warning systems can be such things as barricades, hand or mechanical signals, or stop logs. If possible, the grade should be away from the excavation.

Figure 36. An open excavation with a ladder for egress

Hazardous Atmospheres

To prevent exposure to harmful levels of atmospheric contaminants, and to assure acceptable atmospheric conditions, the following requirements apply:

1. Where oxygen deficiency (atmospheres containing less than 19.5 percent oxygen), or ahazardous atmosphere exists, or could reasonably be expected to exist, such as in excavations in landfill areas, or excavations in areas where hazardous substances are stored nearby, the atmospheres in the excavation is to be tested before employees enter excavations greater than 4 feet in depth.

2. Adequate precautions must be taken to prevent employee exposure to atmospheres containing less than 19.5 percent oxygen, and other hazardous atmospheres.

3. Adequate precaution is to be taken, such as providing ventilation to prevent employee exposure to the atmosphere containing a concentration of a flammable gas in excess of 20 percent of the lower flammable limit of the gas.

4. When controls are used that are intended to reduce the level of atmospheric contaminants to acceptable levels, testing shall be conducted as often as necessary to ensure that the atmosphere remains safe.

Emergency rescue equipment, such as breathing apparatus, a safety harness and line, or a basket stretcher, must be readily available where hazardous atmospheric conditions exist, or may reasonably be expected to develop during work in an excavation. This equipment must be attended when in use.

Employees entering bell-bottom pier holes, or other similar deep and confined footing excavations, must wear a harness with a lifeline securely attached to it. The lifeline is to be separate from any line used to handle materials, and is to be individually attended at all times while the employee, wearing the lifeline, is in the excavation.

Water Accumulation

Employees are not to work in excavations in which there is accumulated water, or in excavations in which water is accumulating, unless adequate precautions have been taken to protect employees against the hazards posed by water accumulation. The precautions necessary to adequately protect employees vary with each situation, but could include special support or shield systems to protect from cave-ins, water removal to control the level of accumulating water, or the use of a safety harness and lifeline.

If water is controlled, or prevented from accumulating by using water removal equipment, the water removal equipment and operations are to be monitored by a competent person to ensure proper operation.

If excavation work interrupts the natural drainage of surface water (such as streams), diversion ditches, dikes, or other suitable means are to be used to prevent surface water from entering the excavation. Adequate drainage of the area adjacent to the excavation must also be provided. Excavations subject to runoff from heavy rains are required to be inspected by a competent person .

Below Level Excavations

Where the stability of adjoining buildings, walls, or other structures are endangered by excavation operations, support systems such as shoring, bracing, or underpinning must be provided for the protection of the employees and to ensure the stability of such structures. Excavation below the level of the base, or footing of any foundation, or retaining walls that could be reasonably expected to pose a hazard to employees, must not be permitted except when

1.   A support system, such as underpinning, is provided to ensure the safety of employees and the stability of the structure.

2.   The excavation is in stable rock.

3.   A registered professional engineer has approved and determined that the structure is sufficiently removed from the excavation so as to be unaffected by the excavation activity.

4.   A registered professional engineer has approved and determined that such excavation work will not pose a hazard to employees.

Sidewalks, pavements, and appurtenant structures are not to be undermined unless a support system or other method of protection is provided to protect employees from the possible collapse of such structures.

Loose Materials

Adequate protection must be provided to protect employees from loose rock or soil that could pose a hazard by falling or rolling from an excavation face. Such protection consists of scaling to remove loose material; installation of protective barricades at intervals, as necessary on the face, to stop and contain falling material; or other means that provide equivalent protection. Employees are to be protected from excavated or other materials or equipment that could pose a hazard by falling or rolling into excavations. Protection shall be provided by placing and keeping such materials or equipment at least 2 feet from the edge of excavations, by using retaining devices that are sufficient to prevent materials or equipment from falling or rolling into excavations, or by a combination of both, if necessary.

Inspections

Daily inspections of excavations, the adjacent areas, and protective systems must be made by a competent person for any evidence of a situation that could result in possible cave-ins, failure of protective systems, hazardous atmospheres, or other hazardous conditions. An inspection is to be conducted by the competent person prior to the start of work, and as needed throughout the shift. Inspections shall also be made after every rainstorm or other hazard which may increase the occurrence of a hazardous condition. These inspections are only required when employee exposure can be reasonably anticipated. When the competent person finds evidence of a situation that could result in a possible cave-in, indication of possible failure of protective systems, hazardous atmospheres, or other hazardous conditions, exposed employees are to be removed from the hazardous area until the necessary precautions have been taken to ensure their safety.

Walkways and Barriers

Walkways are to be provided where employees or equipment are required, or permitted to cross over excavations. Guardrails need to be provided where walkways are 6 feet or more above lower levels. Adequate physical barrier protection must be provided at all remotely located excavations. All wells, pits, shafts, etc. must be barricaded or covered. Upon completion of exploration and other similar operations, temporary wells, pits, shafts, etc. are to be backfilled.

Requirements for Protective Systems 1926.652)

Each employee in an excavation must be protected from cave-ins by an adequate protective system except when

1. Excavations are made entirely in stable rock.

2. Excavations are less than 5 feet in depth and examination of the ground, by a competent person, provides no indication of a potential cave-in.

Protective systems must have the capacity to resist, without failure, all loads that are intended, or could reasonably be expected to be applied or transmitted to the system.

Slopes

The slopes and configurations of sloping and benching systems are to be selected and constructed by the employer or his designee, and must be in accordance with the requirements as follows:

1. Excavations must be sloped at an angle not steeper than one and one-half horizontal to one vertical (34 degrees measured from the horizontal), unless the employer uses one of the other options listed below (see Figure 37).

2. Slopes specified are to be excavated to form configurations that are in accordance with the slopes shown for Type C soil in Appendix B of 29 CFR 1926, Subpart P.

3. Maximum allowable slopes, and allowable configurations for sloping and benching systems, are to be determined in accordance with the conditions and requirements set forth in Appendices A and B of this Subpart P.

4. Designs of sloping or benching systems shall be selected from, and in accordance with tabulated data, such as tables and charts. The tabulated data is to be in written form and shall include the identification of the parameters that affect the selection of a sloping or benching system drawn from such data. The identification of the limits of use of the data must include the magnitude and con-

Figure 37. Example of sloping and excavation

figuration of the slopes determined to be safe, any explanatory information, as may be necessary, to aid the user in making a correct selection of a protective system from the data, and at least one copy of the tabulated data which identifies the registered professional engineer who approved the data. This information must be maintained at the jobsite during construction of the protective system. After that time, the data may be stored off the jobsite.

5.   Sloping and benching systems must be designed by a registered professional engineer. The designs shall be in written form and shall include at least the magnitude of the slopes that were determined to be safe for the particular project, the configurations that were determined to be safe for the particular project, and the identity of the registered professional engineer approving the design.

Support, Shield, and Other Protective Systems

Designs of support systems, shield systems, and other protective systems are selected and constructed by the employer or his designee as follows:

1.   Designs using Appendices A, C, and D. Designs for timber shoring in trenches are determined in accordance with the conditions and requirements set forth in Appendices A and C of Subpart P (see Figure 38). Designs for aluminum hydraulic shoring are to be in accordance with Appendix D of Subpart P.

2.   Design of support systems, shield systems, or other protective systems that are drawn from manufacturer's tabulated data, are to be in accordance with all specifications, recommendations, and limitations issued or made by the manufacturer. Deviation from the specifications, recommendations, and limitations issued, or made by the manufacturer, is only allowed after the manufacturer issues specific written approval.

3.   Manufacturer's specifications, recommendations, and limitations, and manufacturer's approval to deviate from the specifications, recommendations, and limitations must be in written form at the jobsite during construction of the protective system. After that time this data may be stored off the jobsite.

4.   Designs of support systems, shield systems, or other protective systems are selected from and are to be in accordance with tabulated data, such as tables and charts. The tabulated data must be in written form and include identifica-

Figure 38. A timber shored wall of an excavation

tion of the parameters that affect the selection of a protective system drawn from such data, identification of the limits of use of the data, and explanatory information, as may be necessary, to aid the user in making a correct selection of a protective system from the data. At least one copy of the tabulated data, which identifies the registered professional engineer who approved the data, must be maintained at the jobsite during construction of the protective system.

5.   The systems must be designed by a registered professional engineer.

Materials and equipment used for protective systems are to be free from damage or defects that might impair their proper function. Manufactured materials and equipment used for protective systems are to be used and maintained in a manner that is consistent with the recommendations of the manufacturer, and in a manner that will prevent employee exposure to hazards.

When material or equipment that is used for protective systems is damaged, a competent person must examine the material or equipment and evaluate its suitability for continued use. If the competent person cannot assure the material or equipment is able to support the intended loads, or is otherwise suitable for safe use, then such material or equipment shall be removed from service, and shall be evaluated and approved by a registered professional engineer before being returned to service.

Members of support systems are to be securely connected together to prevent sliding, falling, kickouts, or other predictable failure. Support systems must be installed and removed in a manner that protects employees from cave-ins, structural collapses, or from being struck by members of the support system. Individual members of support systems are not to be subjected to loads exceeding those which those members were designed to withstand.

Before temporary removal of individual members begins, additional precautions must be taken to ensure the safety of employees, such as installing other structural members to carry the loads imposed on the support system. Removal begins at, and progresses from the bottom of the excavation. Members are released slowly so as to note any indication of possible failure of the remaining members of the structure, or possible cave-in of the sides of the excavation. Backfilling is to progress together with the removal of support systems from excavations.

Excavation of material, to a level no greater than 2 feet below the bottom of the members of a support system, is permitted, but only if the system is designed to resist the forces calculated for the full depth of the trench, and if there are no indications, while the trench is open, of a possible loss of soil from behind or below the bottom of the support system.

Installation of any support system is to be closely coordinated with the excavation of trenches. Employees are never permitted to work on the faces of sloped or benched excavations at levels above other employees, except when employees at the lower levels are adequately protected from the hazard of falling, rolling, or sliding material or equipment.

Shield Systems

Shield systems must not be subjected to loads exceeding those which the system was designed to withstand. Shields are to be installed in a manner which restricts lateral, or other hazardous movement of the shield, in the event of the application of sudden lateral loads. Employees are to be protected from the hazard of cave-ins when entering or exiting the areas protected by shields. Employees are not allowed in shields when shields are being installed, removed, or moved vertically. Some additional requirements for the use of shields in trenches or excavations are when earth material reaches a level not greater than 2 feet below the bottom of a shield are permitted, only if the shield is designed to resist the forces calculated for the full depth of the trench, and where there are no indications, while the trench is open, of a possible loss of soil from behind or below the bottom of the shield (see Figure 39).

Figure 39. Worker using a trench box as a shield with a ladder for egress

EXPLOSIVES AND BLASTING (1926.900)

Although there are not as many tragedies as in the past from working with explosives, the dangers still exist, and from time to time explosive/blasting related accidents are reported. Much of the blasting which now occurs is carried out by contract blasters who specialize in the use and handling of explosives. All employers, workers, and contract blasters are required to follow the regulatory requirements for use of explosives.

Blaster Qualifications (1926.901)

To begin with, only authorized and qualified persons are permitted to handle and use explosives. These individuals must be able to understand and give written and oral orders. Blasters are to be in good physical condition and not be addicted to narcotics, intoxicants, or similar types of drugs. They must be qualified, by reason of training, knowledge, or experience, in the field of transporting, storing, handling, and using explosives, and have a working knowledge of state and local laws and regulations which pertain to explosives. Blasters are required to furnish satisfactory evidence of competency in the handling explosives, and performing in a safe manner the type of blasting that will be required. They are also required to be knowledgeable and competent in the use of each type of blasting method used.

General Provisions

Smoking, firearms, matches, open flame lamps, other fires, flame or heat producing devices and sparks, are prohibited in or near explosive magazines, or while explosives are being handled, transported, or used. No person is allowed to handle or use explosives while under the influence of intoxicating liquors, narcotics, or other dangerous drugs.

All explosives must be accounted for at all times. Explosives not being used are to be kept in a locked magazine, unavailable to persons not authorized to handle them. The employer shall maintain an inventory and have records of all explosives. Appropriate authorities must be notified of any loss, theft, or unauthorized entry into a magazine. In no case are explosives or blasting agents abandoned.

Employees authorized to prepare explosive charges or conduct blasting operations, must use every reasonable precaution including, but not limited to, visual and audible warning signals, flags, or barricades, to ensure employee safety. When blasting is done in congested areas, or in proximity to a structure, railway, highway, or any other installation that may be damaged, the blaster shall take special precautions in the loading, delaying, initiation, and confinement of each blast by using mats (see Figure 40), or other methods so as to control the throw of fragments; this will prevent bodily injury to employees.

Figure 40. Example of blasting mats

Insofar as possible, blasting operations above ground are to be conducted between sunup and sundown. Precautions are to be taken to prevent accidental discharge of electric blasting caps from current induced by radar, radio transmitters, lightning, adjacent powerlines, dust storms, or other sources of extraneous electricity. These precautions shall include

1. Detonators are to be short-circuited in holes which have been primed and shunted until wired into the blasting circuit.

2. The suspension of all blasting operations, and removal of persons from the blasting area during the approach and progress of an electric storm.

3. The prominent display of adequate signs, warning against the use of mobile radio transmitters on all roads within 1,000 feet of blasting operations. Whenever adherence to the 1,000-foot distance would create an operational handicap, a competent person must be consulted to evaluate the particular situation, and alternative provisions may be made which are adequately designed to prevent any premature firing of electric blasting caps. A description of any such alternatives must be in writing and certified by the competent person consulted, as meeting the purposes of this subdivision. The description shall be maintained at the construction site during the duration of the work, and shall be available for inspection by representatives of the Secretary of Labor.

Blasting operations, in the proximity of overhead power lines, communication lines, utility services, or other services and structures, are not to be carried on until the operators and/ or owners have been notified and measures have been taken for safe control. Cables in the proximity of the blast area must be deenergized and locked out by the blaster.

Great care needs to be taken to ensure that mobile radio transmitters, which are less than 100 feet away from electric blasting caps and are in other than original containers, are deenergized and effectively locked. There must be compliance with the recommendations of The Institute of the Makers of Explosives with regard to blasting in the vicinity of radio transmitters and as stipulated in Radio Frequency Energy-A, Potential Hazard in the Use of Electric Blasting Caps, IME Publication No. 20, March 1971.

Transporting Explosives (1926.902)

When moving explosives, only the original containers or Class II magazines are to be used for taking detonators and other explosives from storage magazines to the blasting area. Transportation of explosives must meet the provisions of the Department of Transportation regulations, contained in 46 CFR Parts 146-149, Water Carriers; 49 CFR Parts 171-179, Highways and Railways; 49 CFR Part 195, Pipelines; and 49 CFR Parts 390-397, Motor Carriers. Motor vehicles or conveyances transporting explosives shall only be driven by, and be in the charge of, a licensed driver who is physically fit. Drivers are to be familiar with the local, State, and Federal regulations governing the transportation of explosives. No person can smoke, carry matches, or any other flame-producing device, nor shall firearms or loaded cartridges be carried while in or near a motor vehicle or conveyance transporting explosives. Explosives, blasting agents, and blasting supplies are not to be transported with other materials or cargoes. Blasting caps (including electric) shall not be transported in the same vehicle with other explosives. Vehicles used for transporting explosives must be strong enough to carry the load without difficulty, and must be in good mechanical condition. When explosives are transported by a vehicle with an open body, a Class II magazine or original manufacturer's container is to be securely mounted on the bed to contain the cargo. All vehicles used for the transportation of explosives shall have tight floors, and any exposed spark-producing metal on the inside of the body shall be covered with wood, or other nonsparking material, to prevent contact with containers of explosives.

Every motor vehicle or conveyance used for transporting explosives shall be marked or placarded on both sides, the front and the rear, with the word "Explosives" in red letters on a white background, and not less than 4 inches in height (see Figure 41). In addition to such marking or placarding, the motor vehicle or conveyance may display, in such a manner that it will be readily visible from all directions, a red flag 18 inches by 30 inches, with the word "Explosives" painted, stamped, or sewed thereon, in white letters, and at least six inches in

Figure 41. Typical vehicle used by those transporting explosives

height. Each vehicle used for transportation of explosives is to be equipped with a fully charged fire extinguisher, in good condition. An Underwriters Laboratory-approved extinguisher of not less than 10-ABC rating will meet the minimum requirement. The driver must be trained in the use of the extinguisher on his or her vehicle. No fire is to be fought where the fire is in imminent danger of contact with explosives. All employees are to be removed to a safe area and the fire area guarded against intruders.

Motor vehicles or conveyances carrying explosives, blasting agents, or blasting supplies, are not to be taken inside a garage or shop for repairs or servicing. No motor vehicle transporting explosives shall be left unattended.

Use of Explosives (1926.904 and 905)

Explosives and related materials are to be stored in approved facilities required under the applicable provisions of the Bureau of Alcohol, Tobacco, and Firearms regulations contained in 27 CFR part 55. Blasting caps, electric blasting caps, detonating primers, and primed cartridges are not stored in the same magazine with other explosives or blasting agents. Smoking and open flames are not permitted within 50 feet of explosives and detonator storage magazine.

The loading of explosives and blasting agents must follow procedures that permit safe and efficient loading before loading is started. All drill holes must be sufficiently large enough to admit freely the insertion of the cartridges of explosives. Tamping is done only with wood rods or plastic tamping poles without exposed metal parts, but non-sparking metal connectors may be used for jointed poles. Violent tamping must be avoided. The primer is never tamped. Holes are not loaded except those to be fired in the next round of blasting. After loading, all remaining explosives and detonators are immediately returned to an authorized magazine.

New drilling is not started until all remaining butts of old holes are examined for unexploded charges, and if any are found, they are refired before work proceeds. No person is allowed to deepen drill holes which have contained explosives or blasting agents. Machines and all tools not used for loading explosives into bore holes are removed from the immediate location of holes before explosives are delivered. No equipment is to be operated within 50 feet of loaded holes. No activity of any nature, other than that which is required for loading holes with explosives, is permitted in a blast area (see Figure 42).

Figure 42. Drilled blasting area ready for loading

Holes are checked prior to loading to determine depth and conditions. Where a hole has been loaded with explosives but the explosives have failed to detonate, there shall be no drilling within 50 feet of the hole. When loading a long line of holes with more than one loading crew, the crews must be separated by a practical distance consistent with efficient operation and supervision of crews. All blast holes in open work are to be stemmed to the collar or to a point which will confine the charge. Warning signs, indicating a blast area, are to be maintained at all approaches to the blast area. The warning sign lettering shall not be less than 4 inches in height and be on a contrasting background. A bore hole is never sprung when it is adjacent to or near a hole that is loaded. Flashlight batteries are not used for springing holes. Drill holes which have been sprung or chambered, and which are not water-filled, must be allowed to cool before explosives are loaded. No loaded holes shall be left unattended or unprotected. When loading blasting agents pneumatically over electric blasting caps, a semiconductive delivery hose is to be used, and the equipment is to be bonded and grounded.

Electrical Blasting (1926.906)

When initiating a blast, electric blasting caps are not to be used where sources of extraneous electricity make the use of electric blasting caps dangerous. Blasting cap leg wires shall be kept short-circuited (shunted) until they are connected into the circuit for firing. Before adopting any system of electrical firing, the blaster is to conduct a thorough survey for extraneous currents, and all dangerous currents shall be eliminated before any holes are loaded. In any single blast using electric blasting caps, all caps are to be of the same style or function, and the same manufacturer. Electric blasting must be carried out by using blasting circuits or power circuits in accordance with the electric blasting cap manufacturer's recommendations, or by an approved contractor or his designated representative. When firing a circuit of electric blasting caps, care must be exercised to ensure that an adequate quantity of delivered current is available, in accordance with the manufacturer's recommendations. Connecting wires and lead wires are to be insulated single solid wires of sufficient current-carrying capacity. Also, bus wires must be solid single wires of sufficient current-carrying capacity. When firing electrically, the insulation on all firing lines is to be adequate and in good condition. The power circuit used for firing electric blasting caps must not be grounded. When firing from a power circuit, the firing switch must be locked in the open or "Off" position at all times, except when firing. It must be so designed that the firing lines to the cap circuit are automatically short-circuited when the switch is in the "Off" position. Keys to this switch are to be entrusted only to the blaster.

Blasting machines in use must be in good condition and the efficiency of the machine tested periodically to make certain that it can deliver power at its rated capacity. When firing with blasting machines, the connections are to be made as recommended by the manufacturer of the electric blasting caps used. The number of electric blasting caps connected to a blasting machine are not to be in excess of its rated capacity. Furthermore, in primary blasting, a series circuit shall contain no more caps than the limits recommended by the manufacturer of the electric blasting caps in use. The blaster is in charge of the blasting machines, and no other person shall connect the leading wires to the machine. Whenever the possibility exists that a leading line or blasting wire might be thrown over a live powerline by the force of an explosion, care must be taken to see that the total length of wires are kept too short to hit the lines, or that the wires are securely anchored to the ground. If neither of these requirements can be satisfied, a non-electric system is to be used. In electrical firing, only the person making leading wire connections shall fire the shot. All connections are to be made from the bore hole back to the source of firing current, and the leading wires must remain shorted and not connected to the blasting machine or other source of current until the charge is to be fired. After firing an electric blast from a blasting machine, the leading wires must immediately be disconnected from the machine and short-circuited.

Safety Fuse (1926.907)

At times when extraneous electricity makes the use of electric blasting caps dangerous, safety fuses are to be used. The use of a fuse that has been hammered or injured in any way is forbidden. The hanging of a fuse on nails or other projections which will cause a sharp bend to be formed in the fuse is prohibited. Before capping a safety fuse, a short length is to be cut from the end of the supply reel so as to assure a fresh cut end in each blasting cap. Only a cap crimper of approved design is to be used for attaching blasting caps to a safety fuse. No unused cap or short capped fuse shall be placed in any hole to be blasted; such unused detonators are removed from the working place and destroyed. No fuse is capped, or primers made up, in any magazine or near any possible source of ignition, and no one is permitted to carry detonators or primers of any kind on his person. The minimum length of a safety fuse to be used in blasting shall not be less than 30 inches. At least two persons must be present when multiple cap and fuse blasting is done by the hand-lighting methods. Not more than 12 fuses are to be lighted by each blaster when hand-lighting devices are used. However, when two or more safety fuses in a group are lighted as one, by means of igniter cord or other similar fuse-lighting devices, they may be considered as one fuse. The so-called "drop fuse" method of dropping or pushing a primer or any explosive with a lighted fuse attached is forbidden. Caps and fuses are not used for firing mudcap charges unless charges are separated sufficiently to prevent one charge from dislodging other shots in the blast. When blasting with safety fuses, consideration must be given to the length and burning rate of the fuse. Sufficient time, with a margin of safety, must always be provided for the blaster to reach a place of safety.

Using Detonating Cord (1926.908)

When a detonating cord is used care must be taken to select a detonating cord consistent with the type and physical condition of the bore hole and stemming, and the type of explosives used. A detonating cord is to be handled and used with the same respect and care given other explosives. The line of a detonating cord, extending out of a bore hole or from a charge, is to be cut from the supply spool before loading the remainder of the bore hole, or placing additional charges. Detonating cord connections are to be competent and positive in accordance with the approved and recommended methods. Knot-type, or other cord-to-cord connections, are to be made only with a detonating cord in which the explosive core is dry. All detonating cord trunklines and branchlines are to be free of loops, sharp kinks, or angles that direct the cord back toward the oncoming line of detonation. All detonating cord connections must be inspected before firing the blast. When detonating cord millisecond-delay connectors or short-interval-delay electric blasting caps are used with a detonating cord, the practice shall conform strictly to the manufacturer's recommendations. When connecting a blasting cap or an electric blasting cap to a detonating cord, the cap is to be taped, or otherwise attached securely along the side or end of the detonating cord, with the end of the cap containing the explosive charge pointed in the direction in which the detonation is to proceed. Detonators for firing the trunkline must not brought to the loading area, nor attached to the detonating cord, until everything else is in readiness for the blast. A code of blasting signals equivalent to Table 6, are to be posted in one or more conspicuous places at the operation, and all employees are required to familiarize themselves with the code and conform to it. Danger signs must be placed at suitable locations.

TABLE 6

Blasting Signals

Courtesy of OSHA

WARNING SIGNAL	– A 1-minute series of long blasts 5 minutes prior to blast signal.
BLAST SIGNAL	– A series of short blasts 1 minute prior to the shot.
ALL CLEAR SIGNAL	– A prolonged blast following the inspection of the blast area.

Firing a Blast (1926.909)

Before a blast is fired, a loud warning signal is given by the blaster in charge, who has made certain that all surplus explosives are in a safe place and all employees, vehicles, and equipment are at a safe distance, or under sufficient cover. Flagmen are to be safely stationed on highways which pass through the danger zone, so as to stop traffic during blasting operations. It is the duty of the blaster to fix the time of blasting. Before firing an underground blast, warning must be given, and all possible entries into the blasting area and any entrances to any working place where a drift, raise, or other opening is about to hole through, must be carefully guarded. The blaster shall make sure that all employees are out of the blast area before firing a blast. Immediately after the blast has been fired, the firing line is to be disconnected from the blasting machine, or where power switches are used, they are to be locked open or in the off position.

Sufficient time must be allowed, not less than 15 minutes in tunnels, for the smoke and fumes to leave the blasted area before returning to the shot. Before employees are allowed to return to the operation, an inspection of the area and the surrounding rubble must be made by the blaster to determine if all charges have been exploded, and an inspection must be made in the tunnels, after the muck pile has been wetted down.

Handling Misfires (1926.911)

If a misfire is found, the blaster must provide proper safeguards for excluding all employees from the danger zone. No other work shall be done except that necessary to remove the hazard of the misfire, and only those employees necessary to do the work shall remain in the danger zone. No attempts are to be made to extract explosives from any charged or misfired hole; a new primer shall be put in and the hole reblasted. If refiring of the misfired hole presents a hazard, the explosives may be removed by washing out with water or, where the misfire is under water, blown out with air. If there are any misfires while using a cap and fuse, all employees must remain away from the charge for at least 1 hour. Misfires shall be handled under the direction of the person in charge of the blasting. All wires are to be carefully traced and a search made for unexploded charges. No drilling, digging, or picking is permitted until all missed holes have been detonated or the authorized representative has approved that work can proceed.

General Guidelines

Empty boxes, paper, and fiber packing materials which have previously contained high explosives, are not to be used again for any purpose; they are to be destroyed by burning at an approved location. Explosives, blasting agents, and blasting supplies that are obviously deteriorated or damaged are not to be used. The use of black powder is prohibited. All loading and firing shall be directed and supervised by competent persons who are thoroughly experienced in this field. Buildings used for the mixing of blasting agents must conform to the requirements of this section. Buildings are to be constructed of noncombustible construction, or sheet metal on wood studs. Floors in a mixing plant must be concrete or of other nonabsorbent materials. All fuel oil storage facilities must be separated from the mixing plant and located in such a manner, that in case of tank rupture, the oil will drain away from the mixing plant building. The building is to be well ventilated. Heating units which do not depend on combustion processes, when properly designed and located, may be used in the building. All direct sources of heat must be provided exclusively from units located outside the mixing building. All internal-combustion engines used for electric power generation are to be located outside the mixing plant building, or must be properly ventilated and isolated by a firewall. The exhaust systems on all such engines are to be located so that any spark emission will not be a hazard to any materials in, or adjacent to the plant.

In summary, only authorized and qualified individuals are permitted to handle and use explosives. Smoking, firearms, matches, open flame lamps, other fires, flames, heat producing devices, and sparks are prohibited in or near explosive magazines. Explosives not being used must be kept in a locked magazine. The blasters must maintain an up-to-date record of explosives, blasting agents, and blasting supplies used and stored on the jobsite. Appropriate authorities must be notified of any loss, theft, or unauthorized entry into a magazine.

EYE AND FACE PROTECTION (1926.102)

If a potential risk of eye or face injury exists from machines or operations, appropriate eye and face protection must be provided. All eye and face protection must meet the requirements of ANSI Z87.1 – 1968, Practices for Occupational and Industrial Eye and Face Protection. If a worker has to wear prescription lenses, the worker needs to use safety approved prescription glasses or goggles which cover, or are incorporated within them, the prescriptive lenses.

Figure 43. Worker using a chop saw with eye and hearing protection

For selection of appropriate protective eye and face wear consult Table E-1 in 29 CFR 1926.102. Eye and face protection should provide the worker with

1.   Adequate protection against the hazard.

2.   Proper fit and comfort.

3.   Durability.

4.   The capability of being disinfected and cleaned.

Welding workers will need to be furnished with the proper shade of filter lenses for a specific welding process; consult Table E-2 in 29 CFR 1926.102 for the filter lens selection. When using laser safety goggles for eye protection from laser beams, consult Table E-3 in 29 CFR 1926,102 for selection of laser protective eyewear (see Figure 43).

FALL PROTECTION (1926.500 - 503)

Scope, Application, and Definitions Applicable to this Subpart (1926.500)

Falls on construction sites are the leading cause of death to construction workers; thus the need for fall protection in construction workplaces. Fall protection is not required when workers are making an inspection, investigation, or assessment of workplace conditions prior to the actual start of construction work, or after all construction work has been completed. Requirements relating to fall protection, as described in this section, does not apply to scaffolds, cranes and derricks, steel erection, ladders and stairways, and tunneling operations, or to electrical power transmission and distribution construction, each of which has its own requirements. The major components of fall protection, described herein, are for installation, construction, and the proper use of body belts, lanyards, and lifelines, and the requirements for the training of fall protection. Some general do's and don'ts for climbing and working at heights are found in Table 7 and Table 8.

Duty to have Fall Protection (1926.501)

It is the employers responsibility to determine if the walking/working surfaces on which its employees are to work have the strength and structural integrity to support employees safely. Employees are allowed to work on those surfaces only when the surfaces have the requisite strength and structural integrity. Any time a worker is on a walking/working surface (horizontal and vertical surface), or constructing a leading edge with an unprotected side or edge which is 6 feet or more above a lower level, the worker must be protected from falling by using guardrail systems, safety net systems, or personal fall arrest systems. If the employer can demonstrate that it is not feasible or creates a greater hazard to use these systems, the employer shall develop and implement a fall protection plan (consult Appendix E in Subpart M of 29 CFR 1926).

Each worker on a walking/working surface 6 feet or more above a lower level where leading edges are under construction, but where the worker is not engaged in the leading edge work, must be protected from falling by using a guardrail system, safety net system, or personal fall arrest system. If a guardrail system is chosen to provide the fall protection, and a controlled access zone has already been established for leading edge work, the control line may be used in lieu of a guardrail along the edge that parallels the leading edge.

Workers in a hoist area are to be protected from falling 6 feet or more to lower levels by using guardrail systems or personal fall arrest systems. If guardrail systems, (chains, gates,

Table 7

Dos for Working at Heights

- Do Close and Latch All Hatches, Security Gates, and Hinged Walkways, to Seal Openings Where There Is Fall Potential.
- Do Barricade or Fence Around Chimneys or Stacks When Working on These Structures.
- Do Tie Off to a Secure Anchor. Ask Yourself if You Would Hang Your New Pickup Truck From This Anchor Point.
- Do Inspect All Fall Protection Equipment Prior to Use.
- Do Wear All Other Types of Personal Protective Equipment Such As Hardhats, Gloves, Hearing Protection, Rain Suits, Respirators, or Protective Eyewear.
- Do Maintain Three Points of Contact at All Times While Climbing (Two Hand and One Foot, or Two Feet and One Hand).
- Do Work Cautiously and Slowly.
- Do Follow All Safety and Health Procedures for Working at Heights.
- Do Wear a Pair of Leather Gloves While Climbing.

Table 8

Don'ts for Working at Heights

- Don't Climb Without Fall Protection.
- Don't Overlook Potential Hazards.
- Don't Climb or Work at Heights During Adverse Weather.
- Don't Assume Someone Else Has Assured Your Safety.
- Don't Wear Jewelry Which Could Catch On Other Objects.
- Don't Climb or Work at Heights if You Evaluate the Conditions as Unsafe.
- Don't Use Any Fall Protection Which Is Worn or Has Not Been Inspected.
- Don't Walk on Roofs Unless You Can Verify They Are Strong Enough to Support You.
- Don't Work at Heights Where There Is a Chance of Falling Without Tying Off.
- Don't Carry Tools or Other Objects, Which Could Slip or Fall, While Tucked Into Your Safety Belt or Harness.
- Don't Hoist Heavy Loads to Your Work Area Without Being Tied Off.
- Don't Climb or Work on Faulty Built Scaffolds.
- Don't Climb or Work on Ladders Which Are Not on Stable Ground, Chocked, or Secure at Base and Secured at the Top.
- Don't Climb a Stack, Chimney, Etc. Unless There Is a Securely Anchored Ladder or Rungs.
- Don't Use Safety Belts, Use Only Full Body Harnesses.
- Don't' Carry Any Loads or Objects in Your Hands as You Climb.

or guardrails), or portions thereof, are removed to facilitate the hoisting operation (e.g., during landing of materials), and the worker must lean through the access opening, or out over the edge of the access opening (to receive or guide equipment and materials, for example), that employee must be protected from fall hazards by using a personal fall arrest system (see Figure 44).

Figure 44. Worker with personal fall protection in a hoist area

Each worker on a walking/working surface is to be protected from tripping in, or stepping into or through holes (including skylights), and from objects falling through holes (including skylights), by using covers.

Each employee on the face of formwork or reinforcing steel shall be protected from falling 6 feet or more to lower levels by using personal fall arrest systems, safety net systems, or positioning device systems. Also, workers on ramps, runways, and other walkways are to be protected from falling 6 feet or more to lower levels by using guardrail systems; workers at the edge of an excavation 6 feet or more in depth are to be protected from falling by using guardrail systems, fences, or barricades; and, workers at the edge of a well, pit, shaft, and similar excavation which is 6 feet (1.8 m) or more in depth must be protected from falling by using guardrail systems, fences, barricades, or covers.

Workers less than 6 feet above dangerous equipment are to be protected from falling into or onto the dangerous equipment by the use of guardrail systems or equipment guards. When workers are 6 feet or more above dangerous equipment, they must be protected from fall hazards by using guardrail systems, personal fall arrest systems, or safety net systems.

When workers are performing overhand bricklaying and related work 6 feet or more above lower levels, they must be protected from falling by using guardrail systems, safety net systems, personal fall arrest systems, or they must work in a controlled access zone. If these workers must reach more than 10 inches below the level of the walking/working surface on which they are working, they are to be protected from falling by using a guardrail system, safety net system, or personal fall arrest system.

Workers engaged in roofing activities on low-slope roofs, with unprotected sides and edges 6 feet or more above lower levels, are to be protected from falling by using guardrail systems, safety net systems, or personal fall arrest systems; or, they are to be protected by using a combination of a warning line system and guardrail system, a warning line system and safety net system, a warning line system and personal fall arrest system, or a warning line system and safety monitoring system. When on a steep roof with unprotected sides and edges 6 feet (1.8 m) or more above lower levels, workers must be protected from falling by using guardrail systems with toeboards, safety net systems, or personal fall arrest systems.

During the erection of precast concrete members (including, but not limited to the erection of wall panels, columns, beams, and floor and roof "tees"), and related operations such as grouting of precast concrete members, those working 6 feet or more above lower levels are to be protected from falling by using guardrail systems, safety net systems, or personal fall arrest systems, unless the employer can demonstrate that it is not feasible, or creates a greater hazard to use these systems. The employer must then develop and implement a fall protection plan. *Note:* There is a presumption that it is feasible and will not create a greater hazard to implement at least one of the above-listed fall protection systems. Accordingly, the employer has the burden of establishing an appropriate fall protection plan.

When workers are engaged in residential construction activities 6 feet or more above lower levels, they must be protected by guardrail systems, safety net systems, or personal fall arrest systems unless the employer can demonstrate that it is not feasible, or creates a greater hazard to use these systems.

Each employee working on, at, above, or near wall openings (including those with chutes attached), where the outside bottom edge of the wall opening is 6 feet or more above lower levels, and the inside bottom edge of the wall opening is less than 39 inches above the walking/working surface, must be protected from falling by using a guardrail system, safety net system, or personal fall arrest system.

When workers are exposed to falling objects, the employer must have each employee wear a hard hat and implement one of the following measures:

1. Erect toeboards, screens, or guardrail systems to prevent objects from falling from higher levels.

2. Erect a canopy structure and keep potential fall objects far enough from the edge of the higher level so that those objects would not go over the edge if they were accidentally displaced (see Figure 45).

3. Barricade the area to which objects could fall; prohibit employees from entering the barricaded area; and keep objects that may fall far enough away from the edge of a higher level so that those objects would not go over the edge if they were accidentally displaced.

Fall Protection Systems Criteria and Practices (1926.502)

Figure 45. Canopy for overhead protection

Employers are to provide and install all fall protection systems before an employee begins the construction work that necessitates the fall protection. The fall protection system, selected by the employer, is to be the one which the employer deems is most appropriate for protecting the workforce.

Guardrail Systems

Guardrail systems are to be composed of the top rail, midrail, and toeboard. The top edge height of top rails, or equivalent guardrail system members, are to be 42 inches, plus or minus 3 inches, above the walking/working level. When conditions warrant, the height of the top edge may exceed the 45-inch height, provided the guardrail system meets all other criteria. *Note:* When employees are using stilts, the top edge height of the top rail, or equivalent member, shall be increased an amount equal to the height of the stilts. Guardrail systems are to be capable of withstanding, without failure, a force of at least 200 pounds applied within 2 inches of the top edge, in any outward or downward direction, at any point along the top edge. When the 200-pound test load is applied in a downward direction, the top edge of the guardrail must not deflect to a height less than 39 inches above the walking/working level. Guardrail system components are to be selected and constructed in accordance with the Appendix B to Subpart M.

Midrails, screens, mesh, intermediate vertical members, or equivalent intermediate structural members are to be installed between the top edge of the guardrail system and the walking/working surface, when there is no wall or parapet wall at least 21 inches high. Midrails, when used, are to be installed at a height midway between the top edge of the guardrail system and the walking/working level. Screens and mesh, when used, must extend from the top rail to the walking/working level, and along the entire opening between the top rail supports. Intermediate members (such as balusters), when used between posts, must not be more than 19 inches apart. Other structural members (such as additional midrails and architectural panels) shall be installed such that there are no openings in the guardrail system that are more than 19 inches (.5 m) wide. Midrails, screens, mesh, intermediate vertical members, solid panels, and equivalent structural members shall be capable of withstanding, without failure, a force of at least 150 pounds applied in any downward or outward direction at any point along the midrail or other member (see Figure 46).

Guardrail systems must be so surfaced as to prevent injury to an employee from punctures or lacerations, and to prevent snagging of clothing. The ends of all top rails and midrails are not to overhang the terminal posts, except where such overhang does not constitute a projection hazard. Steel banding and plastic banding are not to be used as top rails or midrails. Top rails and midrails shall be at least one-quarter inch nominal diameter or thickness to prevent cuts and lacerations. If wire rope is used for top rails, it must be flagged at not more than 6-foot intervals, with high-visibility material. When manila, plastic, or synthetic ropes are being used for top rails or midrails, they are to be inspected as frequently as necessary to ensure that they continue to meet the strength requirements.

When guardrail systems are used at hoisting areas, a chain, gate or removable guardrail section must be placed across the access opening between guardrail sections, when hoisting operations are not taking place. When guardrail systems are used at holes, they are to beerected on all unprotected sides or edges of the hole. When guardrail systems are used around holes, and the guardrails are also used for the passage of materials, the guardrails must not have more than two sides with removable guardrail sections which will be used for the passage of the materials. When the hole is not in use, it must be closed over with a cover, or a guardrail system must be provided along all unprotected sides or edges. When guardrail systems are used around holes which are used as points of access (such as ladderways), they must be

Figure 46. Guardrail system on formwork

provided with a gate, or be so offset that a person cannot walk directly into the hole. Guardrail systems used on ramps and runways are to be erected along each unprotected side or edge.

Safety Nets

Safety nets are to be installed as close as practicable under the walking/working surface on which employees are working, but in no case more than 30 feet below such level. When nets are used on bridges, the potential fall area from the walking/working surface to the net is to be unobstructed. Safety nets shall extend outward from the outermost projection of the work surface as follows:

Vertical distance from working level to horizontal plane of net	Minimum required horizontal distance of outer edge of net from the edge of the working surface
Up to 5 feet	8 feet
More than 5 feet up to 10 feet	10 feet
More than 10 feet	13 feet

Safety nets are to be installed with sufficient clearance under them to prevent contact with the surface or structures below, when subjected to an impact force equal to the drop test. The safety nets and safety net installations are to be drop-tested at the jobsite after the initial installation. When safety nets are relocated, they must be retested (drop-tested) before they are used again as a fall protection system. They must also be drop-tested after a major repair, and at 6-month intervals, if left in one place. The drop-test consists of a 400-pound bag of sand 30

+ or – 2 inches in diameter, dropped into the net from the highest walking/working surface at which employees are exposed to fall hazards, but not from less than 42 inches above that level. When the employer can demonstrate that it is unreasonable to perform the drop-test required, then the employer (or a designated competent person) must certify that the net and net installation is in compliance the strength requirements, by preparing a certification record prior to the net being used as a fall protection system. The certification record must include an identification of the net and net installation for which the certification record is being prepared; the date that it was determined that the identified net and net installation were in compliance; and the signature of the person making the determination and certification. The most recent certification record for each net and net installation is to be available for inspection at the jobsite (see Figure 47).

Figure 47. Example of a perimeter safety net

Defective nets are not to be used. Safety nets must be inspected at least once a week for wear, damage, and other deterioration. Defective components shall be removed from service and safety nets shall also be inspected after any occurrence which could affect the integrity of the safety net system. Materials, scrap pieces, equipment, and tools which have fallen into the safety net are to be removed as soon as possible from the net, and at least before the next work shift.

The maximum size of each safety net mesh opening shall not exceed 36 square inches (230 cm), nor be longer than 6 inches on any side, and the opening, measured center-to-center of mesh ropes or webbing, shall not be longer than 6 inches. All mesh crossings are to be secured to prevent enlargement of the mesh opening. Each safety net (or section of it) must have a border rope, for webbing, with a minimum breaking strength of 5,000 pounds. Connections between safety net panels are to be as strong as integral net components, and must be spaced not more than 6 inches apart.

Personal Fall Arresting System

Effective January 1, 1998, body belts are not acceptable as part of a personal fall arrest system. Note: The use of a body belt in a positioning device system is acceptable. Connectors are to be drop forged, pressed or formed steel, or made of equivalent materials. Connectors must have a corrosion-resistant finish, and all surfaces and edges shall be smooth to prevent damage to interfacing parts of the system. Ropes and straps (webbing) used in lanyards, lifelines, and strength components of body belts and body harnesses are to be made from synthetic fibers (Figure 48).

Dee-rings and snaphooks must have a minimum tensile strength of 5,000 pounds. Dee-rings and snaphooks are to be proof-tested to a minimum tensile load of 3,600 pounds without cracking, breaking, or taking permanent deformation. Snaphooks shall be sized to be compatible with the member to which they are connected in order to prevent unintentional disengagement of the snaphook by depression of the snaphook keeper by the connected member; or, they are to be a locking type snaphook designed and used to prevent disengagement of the snaphook by the contact of the snaphook keeper by the connected member (see Figure 49). Effective January 1, 1998, only locking type snaphooks are to be used. Unless the snaphook is a locking type and designed for the following connections, snaphooks are not to be engaged:

1. Directly to webbing, rope, or wire rope.

2. To each other.

3. To a dee-ring to which another snaphook or other connector is attached.

4. To a horizontal lifeline.

5. To any object which is incompatibly shaped or dimensioned in relation to the snaphook such that unintentional disengagement could occur by the connected object being able to depress the snaphook keeper and release itself.

Figure 48. Personal fall protection system

On suspended scaffolds, or similar work platforms with horizontal lifelines which may become vertical lifelines, the devices used to connect to a horizontal lifeline are to be capable of locking in both directions on the lifeline. Horizontal lifelines must be designed, installed, and used, under the supervision of a qualified person, as part of a complete personal fall arrest system, which maintains a safety factor of at least two. Lanyards and vertical lifelines must have a minimum breaking strength of 5,000 pounds. Except when vertical lifelines

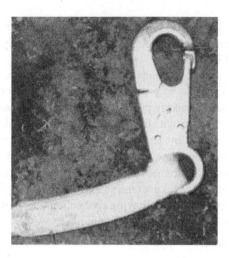

Figure 49. Double locking snaphook

are used, each employee is to be attached to a separate lifeline. During the construction of elevator shafts, two employees may be attached to the same lifeline in the hoistway, provided both employees are working atop a false car that is equipped with guardrails; the strength of the lifeline must be 10,000 pounds [5,000 pounds per employee attached]. Lifelines are to be protected against being cut or abraded.

Self-retracting lifelines and lanyards, which automatically limit free fall distance to 2 feet or less, are to be capable of sustaining a minimum tensile load of 3,000 pounds applied to the device with the lifeline or lanyard in the fully extended position. Self-retracting lifelines, ripstitch lanyards, and tearing and deforming lanyards, where the free fall distance is greater that 2 feet, must be capable of sustaining a minimum tensile load of 5,000 pounds applied to the device with the lifeline or lanyard in the fully extended position.

Anchorages used for attachment of personal fall arrest equipment are to be independent of any anchorage being used to support or suspend platforms, and must be capable of supporting at least 5,000 pounds per employee attached, or designed, installed, and used as follows:

1.   As part of a complete personal fall arrest system which maintains a safety factor of at least two.

2.   Under the supervision of a qualified person.

Any personal fall arrest system that is used to stop a fall must limit the maximum arresting force on an employee to 900 pounds when it is used with a body belt. When it is used with a body harness, limit the maximum arresting force on an employee to 1,800 pounds. The body harness must be rigged such that an employee can neither free fall more than 6 feet (1.8 m), nor contact any lower level, and it must bring an employee to a complete stop and limit the maximum deceleration distance an employee travels to 3.5 feet. The system must also have sufficient strength to withstand twice the potential impact energy of an employee free falling a distance of 6 feet, or the free fall distance permitted by the system, whichever is less. *Note:* If the personal fall arrest system meets the criteria and protocols contained in Appendix C to Subpart M, and if the system is being used by an employee having a combined person and tool weight of less than 310 pounds, the system will be considered to be in compliance. If the system is used by an employee having a combined tool and body weight of 310 pounds or more, then the employer must appropriately modify the criteria and protocols of Appendix C to provide proper protection for such heavier weights or the system will not be deemed to be in compliance

Personal fall arrest systems are to be inspected prior to each use, for wear, damage, and other defects. The attachment point of the body harness is to be located in the center of the wearer's back and near shoulder level, or above the wearer's head. Full body harnesses and components are to be used only for employee protection (as part of a personal fall arrest system or positioning device system), and not to hoist materials. Personal fall arrest systems and components subjected to impact loading must be removed immediately from service and must not be used again for employee protection until inspected and determined, by a competent person, to be undamaged and suitable for reuse. Deteriorated and defective components are to be removed from service.

Personal fall arrest systems must not be attached to guardrail systems, nor shall they be attached to hoists. When a personal fall arrest system is used at hoist areas, it is to be rigged to allow the movement of the employee only as far as the edge of the walking/working surface. In the event of a fall, the employer is to provide for prompt rescue of employees or assure that employees are able to rescue themselves.

Positioning device systems, and their use, must conform to the following: they are to be rigged such that an employee cannot free fall more than 2 feet and they are to be secured to an anchorage capable of supporting at least twice the potential impact load of an employee's fall or 3,000 pounds, whichever is greater. The connectors, dee-rings, snaphooks, inspection, and synthetic webbing have the same requirements as did the full body harness.

Warning Lines

Warning lines are to be erected around all sides of the roof work area. When mechanical equipment is not being used, the warning line must be erected not less than 6 feet from the roof edge, but when mechanical equipment is being used, the warning line is to be erected not less than 6 feet from the roof edge which is parallel to the direction of mechanical equipment operation, and not less than 10 feet from the roof edge which is perpendicular to the direction of mechanical equipment operation.

Points of access, materials handling areas, storage areas, and hoisting areas shall be connected to the work area by an access path formed by two warning lines. When the path to a point of access is not in use, a rope, wire, chain, or other barricade, equivalent in strength and height to the warning line, must be placed across the path at the point where the path intersects the warning line erected around the work area, or the path must be offset such that a person cannot walk directly into the work area.

Warning lines consisting of ropes, wires, chains, and supporting stanchions are to be flagged at not more than 6-foot intervals with high-visibility material and are to be rigged and supported in such a way that its lowest point (including sag) is no less than 34 inches from the walking/working surface; its highest point is to be no more than 39 inches from the walking/ working surface.

Once erected with the rope, wire, or chain attached, stanchions must be capable of resisting, without tipping over, a force of at least 16 pounds applied horizontally against the stanchion. It must be 30 inches above the walking/working surface, perpendicular to the warning line, and in the direction of the floor, roof, or platform edge. The rope, wire, or chain must be a minimum tensile strength of 500 pounds, and after being attached to the stanchions, must be capable of supporting, without breaking, the previously described loads. The line shall be attached at each stanchion in such a way that pulling on one section of the line between stanchions will not result in slack being taken up in adjacent sections before the stanchion tips over.

No worker is allowed in the area between a roof edge and a warning line unless the employee is performing roofing work in that area. Mechanical equipment on roofs is to be used or stored only in areas where employees are protected by a warning line system, guardrail system, or personal fall arrest system.

These controlled access zones are to be areas where leading edge and other operations are taking place and the controlled access zone must be defined by a control line, or by any other means that restricts access. When control lines are used, they are to be erected not less than 6 feet, nor more than 25 feet from the unprotected or leading edge, except when erecting precast concrete members. When erecting precast concrete members, the control line is to be erected not less than 6 feet, more than 60 feet, or half the length of the member being erected, whichever is less, from the leading edge. The control line must extend along the entire length of the unprotected or leading edge, and must be approximately parallel to the unprotected or leading edge. The control line is to be connected, on each side, to a guardrail system or wall. When used to control access to areas where overhand bricklaying and related work is taking place,

1. The controlled access zone is to be defined by a control line erected not less than 10 feet, nor more than 15 feet from the working edge.

2. The control line must extend for a distance sufficient for the controlled access zone to enclose all employees performing overhand bricklaying and related work at the working edge, and is to be approximately parallel to the working edge.

3. Additional control lines are to be erected at each end to enclose the controlled access zone.

4. Only employees engaged in overhand bricklaying or related work are permitted in the controlled access zone.

Control lines consist of ropes, wires, tapes, or equivalent materials, and supporting stanchions for overhand bricklaying and are to be flagged, or otherwise clearly marked, at not more than 6-foot intervals with high-visibility material, and are to be rigged and supported in such a way that its lowest point (including sag) is not less than 39 inches from the walking/working surface. Its highest point shall not be more than 45 inches (50 inches when overhand bricklaying operations are being performed) from the walking/working surface. Each line must have a minimum breaking strength of 200 pounds (see Figure 50). On floors and roofs where guardrail systems are not in place prior to the beginning of overhand bricklaying operations, controlled access zones are to be enlarged, as necessary, to enclose all points of access, material handling areas, and storage areas. On floors and roofs where guardrail systems are in place, but need to be removed to allow overhand bricklaying work or leading edge work to take place, only that portion of the guardrail necessary to accomplish that day's work is to be removed.

Safety Monitoring System

When an employer designates a competent person to monitor the safety of other employees, that person is responsible for monitoring and recognizing fall hazards, and warning the workers when it appears that they are unaware of a fall hazard, or are acting in an unsafe manner. The person monitoring the other workers is to be on the same walking/working surface as the employees being monitored; is to be within visual sighting distance of the employees being monitored; is to be close enough to communicate orally with the employees; and must not have other responsibilities which could take the monitor's attention from the monitoring function.

Figure 50. Control lines on the perimeter of an open floor

Mechanical equipment is not to be used or stored in areas where safety monitoring systems are being used to monitor employees engaged in roofing operations on low-slope roofs. No employee, other than an employee engaged in roofing work (on low-sloped roofs), or an employee covered by a fall protection plan, is allowed in an area where an employee is being protected by a safety monitoring system.

Each employee working in a controlled access zone must comply immediately with any fall hazard warnings from safety monitors. See Table 9 which provides guidance regarding which fall protection is appropriate for different types of construction work.

Covers

Covers are to be used for holes in floors, roofs, and other walking/working surfaces. Covers located in roadways and vehicular aisles shall be capable of supporting, without failure, at least twice the maximum axle load of the largest vehicle expected to cross over the cover. All other covers are to be capable of supporting, without failure, at least twice the weight of the employees, equipment, and materials that may be imposed on the cover at any one time. All covers, when installed, must be secured so as to prevent accidental displacement by the wind, equipment, or employees (see Figure 51). Covers must be color coded or marked with the word "HOLE" or "COVER" to provide warning of the hazard. *Note:* This does not apply to cast iron manhole covers, or steel grates used on streets or roadways.

Overhead Protection

Protection from falling objects can mitigate when toeboards are used Toeboards are to be erected along the edge of an overhead walking/working surface for a distance sufficient to protect employees below. Toeboards shall be capable of withstanding, without failure, a force of at least 50 pounds applied in any downward or outward direction at any point along the toeboard. The minimum vertical height for a toeboard is 3 1/2 inches from the top edge to the level of the walking/working surface, and there must not be more than 1/4 inch clearance above the walking/working surface. Toeboards are to be solid, or have openings not over 1 inch at the greatest dimension. Where tools, equipment, or materials are piled higher than the top edge of a toeboard, paneling or screening is to be erected from the walking/working sur-

Table 9
Fall Protection Safeguards

Types	Guardrails	Safety Nets	Personal Fall Arrest Systems	Other
Open Sided Floors	X	X	X	
Leading Edge	X	X	X	Fall Protection Plan (FPP)
Hoist Areas	X	X	X	
Holes (Roofs, Floors)	X	X	X	Secure Covers Marked Danger
Formwork/Rebar		X	X	> 2ft. Positioning Device
Ramps, Runways, Walkways	X			
Wells, Pits, Shafts	X			Fences, Covers, Barricades
Above Dangerous Equipment	X	X	X	Equipment Guards
Overhand Bricklaying	X	X	X	Controlled Access Zone
Roofing				
Low Pitch	X	X	X	Warning Lines 6ft from edge, guardrails, safety nets, PFAS, Safety Monitors (most common)
Steep Pitch	X	X	X	
Precast	X	X	X	FPP
Residential (24 feet)	X	X	X	Slope < 8 in. 12 in. Safety Monitor, FPP
Wall Openings <39 in. High, > 18 in. Wide	X	X	X	

face or toeboard and to the top of a guardrail system's top rail or midrail, for a distance sufficient to protect workers below.

Guardrail systems, when used for falling object protection, must have all openings small enough to prevent passage of potential falling objects. During the performance of overhand bricklaying and related work, no materials or equipment, except masonry and mortar, are to be stored within 4 feet of the working edge. In order to keep work areas clear, excess mortar, broken or scattered masonry units, and all other materials and debris are to be removed, at regular intervals, from the work areas.

During the performance of roofing work, materials and equipment are not to be stored within 6 feet of a roof edge unless guardrails are erected at the edge, and materials which are piled, grouped, or stacked near a roof edge are stable and self-supporting.

Canopies, when used as falling object protection, are to be strong enough to prevent collapse and prevent penetration by any objects which may fall onto the canopy.

Figure 51. Covered and guarded floor opening (Note: although it cannot be seen, "Hole" was painted on this cover.)

Fall Protection Plan

Employers engaged in leading edge work, precast concrete erection work, or residential construction work, who can demonstrate that it is not feasible, or would create a greater hazard to use conventional fall protection equipment, can elect to develop and implement an alternate fall protection plan, but, the fall protection plan must conform to the following:

1. The fall protection plan is to be prepared by a qualified person and developed specifically for the site where the leading edge work, precast concrete work, or residential construction work is being performed; and, the plan must be maintained up to date.

2. Any changes to the fall protection plan is to be approved by a qualified person.

3. A copy of the fall protection plan, with all approved changes, are to be maintained at the job site.

4. The implementation of the fall protection plan is to be under the supervision of a competent person.

5. The fall protection plan must document the reasons why the use of conventional fall protection systems (guardrail systems, personal fall arrest systems,

or safety nets systems) are not feasible, or why their use would create a greater hazard.

6. The fall protection plan must include a written discussion of other measures that will be taken to reduce or eliminate the fall hazard for workers who cannot be provided with protection from the conventional fall protection systems. For example, the employer must discuss the extent to which scaffolds, ladders, or vehicle-mounted work platforms can be used to provide a safer working surface and thereby reduce the hazard of falling.

7. The fall protection plan must identify each location where conventional fall protection methods cannot be used. These locations shall then be classified as controlled access zones.

8. Where no other alternative measure has been implemented, the employer must implement a safety monitoring system.

9. The fall protection plan must include a statement which provides the name or other method of identification for each employee who is designated to work in controlled access zones. No other employees may enter the controlled access zones.

10. In the event an employee falls, or some other related, serious incident occurs, (e.g., a near miss), the employer must investigate the circumstances of the fall, or other incident, to determine if the fall protection plan needs to be changed (e.g., new practices, procedures, or training), and shall implement those changes to prevent similar types of falls or incidents.

Training Requirements (1926.503)

Employers are required to provide a training program for each employee who might be exposed to fall hazards. The program must enable each worker to recognize the hazards of falling, and be trained in the procedures to be followed in order to minimize these hazards. Each employee must be trained, as necessary, by a competent person qualified in the following areas:

1. The nature of fall hazards in the work area.

2. The correct procedures for erecting, maintaining, disassembling, and inspecting the fall protection systems to be used.

3. The use and operation of guardrail systems, personal fall arrest systems, safety net systems, warning line systems, safety monitoring systems, controlled access zones, and other protection to be used.

4. The role of each employee in the safety monitoring system when this system is used.

5. The limitations on the use of mechanical equipment during the performance of roofing work on low-sloped roofs.

6. The correct procedures for the handling and storage of equipment and materials, and the erection of overhead protection.

7. The role of employees in fall protection plans.

8. The standards contained in this subpart.

The employer must prepare a written certification record. The written certification record is to contain the name or other identity of the employee trained, the date(s) of the training, and the signature of the person who conducted the training, or the signature of the employer. If the employer relies on training conducted by another employer, or training completed prior to the effective date of this section, the certification record must indicate the date the employer determined the prior training was adequate, rather than the date of actual training. The latest training certification is to be maintained.

When an employer has reason to believe that an employee, who has already been trained, does not have the understanding or required skills, the employer shall retrain that employee. Circumstances where retraining is required include, but are not limited to, situations where

1.   Changes in the workplace render previous training obsolete.

2.   Changes in the types of fall protection systems, or equipment to be used, render previous training obsolete.

3.   Inadequacies in an employee's knowledge, or use of fall protection systems or equipment indicate that the employee has not retained the requisite understanding or skill.

FIRE PROTECTION AND PREVENTION (1926.150)

The possibility of fire on a construction site needs to be given careful consideration since the potential exists due to hot work, flammable and combustible materials, and the presence of ignition and fuel sources which are omnipresent. Efforts must be undertaken to prevent the occurrence of fires. See Table 10 for a summary of fire protection and prevention.

The employer must provide a fire protection program, as well as firefighting equipment, and it must be followed throughout all phases of the construction and demolition work. Access to all available firefighting equipment is to be maintained at all times. All firefighting equipment, provided by the employer, is to be conspicuously located. All firefighting equipment must be periodically inspected, and maintained in operating condition. Defective equipment is to be immediately replaced. Depending upon the project, the employer shall have a trained and equipped firefighting organization (Fire Brigade) to assure adequate protection to life.

A temporary or permanent water supply, of sufficient volume, duration, and pressure, which will properly operate the firefighting equipment, must be made available. If underground water mains are to be used for firefighting, they are to be installed, completed, and made available for use as soon as practicable.

A fire extinguisher, rated not less than 2A, is to be provided for each 3,000 square feet of the protected building area, or major fraction thereof. Travel distance from any point of the protected area to the nearest fire extinguisher, shall not exceed 100 feet (see Figure 52). Carbon tetrachloride and other toxic vaporizing liquid fire extinguishers are prohibited. Portable fire extinguishers are to be inspected periodically and maintained in accordance with Maintenance and Use of Portable Fire Extinguishers, NFPA No. 10A-1970.One 55-gallon open drum of water with two fire pails may be substituted for a fire extinguisher. Also, a 1/2-inch diameter garden-type hose line, not to exceed 100 feet in length, equipped with a nozzle with a capacity of 5 gallons per minute, may be substituted for a fire extinguisher, if the hose stream can be applied to all points in the area. Each floor must have one or more fire extinguishers. On multi-story buildings, at least one fire extinguisher must be located adjacent to stairway. Extinguishers and water drums, subject to freezing, are to be protected from freezing.

Table 10

General Rules for Fire Protection and Prevention

Fire Protection

1. Access to all available firefighting equipment will be maintained at all times.
2. Firefighting equipment will be inspected periodically and maintained in operating condition. Defective or exhausted equipment must be replaced immediately.
3. All firefighting equipment will be conspicuously located at each jobsite.
4. Fire extinguishers, rated not less than 2A, will be provided for each 3,000 square feet of the protected work area. Travel distance from any point of the protected area to the nearest fire extinguisher must not exceed 100 feet. One 55-gallon open drum of water, with two fire pails, may be substituted for a fire extinguisher having a 2A rating.
5. Extinguishers and water drums exposed to freezing conditions must be protected from freezing.
6. Do not remove or tamper with fire extinguishers installed on equipment or vehicles, or in other locations, unless authorized to do so or in case of fire. If you use a fire extinguisher, be sure it is recharged or replaced with another fully charged extinguisher.

TYPES OF FIRES

- Class A (wood, paper, trash) - use water or foam extinguisher.
- Class B (flammable liquids, gas, oil, paints, grease) - use foam, CO_2, or dry chemical extinguisher.
- Class C (electrical) - use CO_2 or dry chemical extinguisher.
- Class D (combustible metals) - use dry powder extinguisher only.

Fire Prevention

1. Internal combustion engine powered equipment must be located so that exhausts are away from combustible materials.
2. Smoking is prohibited at, or in the vicinity of operations which constitute a fire hazard. Such operations must be conspicuously posted: "No Smoking or Open Flame."
3. Portable battery powered lighting equipment must be approved for the type of hazardous locations encountered.
4. Combustible materials must be piled no higher than 20 feet. Depending on the stability of the material being piled, this height may be reduced.
5. Keep driveways between and around combustible storage piles at least 15 feet wide and free from accumulation of rubbish, equipment, or other materials.
6. Portable fire extinguishing equipment, suitable for anticipated fire hazards on the jobsite, must be provided at convenient, conspicuously accessible locations.
7. Fire fighting equipment must be kept free from obstacles, equipment, materials and debris that could delay emergency use of such equipment. Familiarize yourself with the location and use of the project's fire fighting equipment.
8. Discard and/or store all oily rags, waste, and similar combustible materials in metal containers on a daily basis.
9. Storage of flammable substances on equipment or vehicles is prohibited unless such unit has adequate storage area designed for such use.

Table 10

General Rules for Fire Protection and Prevention *(continued)*

Flammable and Combustible Liquids

1. Explosive liquids, such as gasoline, shall not be used as cleaning agents. Use only approved cleaning agents.

2. Store gasoline and similar combustible liquids in approved and labeled containers in well-ventilated areas free from heat sources.

3. Handling of all flammable liquids by hand containers must be in approved type safety containers with spring closing covers and flame arrestors.

4. Approved wooden or metal storage cabinets must be labeled in conspicuous lettering: "Flammable-Keep Fire Away."

5. Never store more than 60 gallons of flammable, or 120 gallons of combustible liquids in any one approved storage cabinet.

6. Storage of containers shall not exceed 1,100 gallons in any one pile or area. Separate piles or groups of containers by a 5-foot clearance. Never place a pile or group within 20 feet of a building. A 12-foot wide access way must be provided within 200 feet of each container pile to permit approach of fire control apparatus.

Figure 52. Typical fire extinguisher on a construction site

A fire extinguisher, rated not less than 10B, is to be provided within 50 feet of wherever more than 5 gallons of flammable or combustible liquids, or 5 pounds of flammable gas are being used on the jobsite. This requirement does not apply to the integral fuel tanks of motor vehicles.

Fire hose and connections, with one hundred feet or less of 1 1/2-inch hose, and with a nozzle capable of discharging water at 25 gallons or more per minute, may be substituted, in a designated area, for a fire extinguisher rated not more than 2A. This may be done if the hose line can reach all points in the area. When fire hose connections are not compatible with local firefighting equipment, the contractor must provide adapters, or the equivalent, to permit connections.

If the facility being constructed includes the installation of an automatic sprinkler protection, and the installation closely follows the construction, and is placed in service as soon as applicable, laws permit following completion of each story, and this is considered fixed firefighting equipment.

In all structures where standpipes are required, or where standpipes exist in structures being altered, they shall be brought up as soon as applicable laws permit, and are maintained as construction progresses in such a manner that they are always ready for fire protection use. The standpipes are to be provided with Siamese fire department connections on the outside of the structure, at street level, and be conspicuously marked. There must be at least one standard hose outlet at each floor.

An alarm system (e.g., telephone system, siren, etc.) is to be established by the employer whereby employees on the site, and the local fire department, can be alerted of an emergency. The alarm code and reporting instructions must be conspicuously posted at the phones and employee entrances.

Fire walls and exit stairways, required for the completed buildings, are to be given construction priority. Fire doors, with automatic closing devices, are to be hung on openings, as soon as practicable. Fire cutoffs are to be retained in buildings undergoing alterations or demolition until operations necessitate their removal.

Fire Prevention – Storage (CFR 1926.151)

Internal combustion engine powered equipment shall be so located that the exhausts are well away from combustible materials. When the exhausts are piped to the outside of a building under construction, a clearance of at least 6 inches is to be maintained between such piping and combustible material.

Smoking is prohibited at, or in the vicinity of operations which constitute a fire hazard, and are to be conspicuously posted: "No Smoking or Open Flame."

Portable battery powered lighting equipment, used in connection with the storage, handling, or use of flammable gases or liquids, are to be of the type approved for the hazardous locations.

The nozzle of air, inert gas, and steam lines or hoses, when used in the cleaning or ventilation of tanks and vessels that contain hazardous concentrations of flammable gases or vapors, are to be bonded to the tank or vessel shell. Bonding devices are not to be attached or detached in hazardous concentrations of flammable gases or vapors.

No temporary building shall be erected where it will adversely affect any means of exit. Temporary buildings, when located within another building or structure, must be of either noncombustible construction, or combustible construction having a fire resistance of not less than one hour.

Temporary buildings, located other than inside another building and not used for storage, handling, or use of flammable or combustible liquids, flammable gases, explosives, blasting agents, or similar hazardous occupancies, are to be located at a distance of not less than 10 feet from another building or structure. Groups of temporary buildings, not exceeding 2,000 square feet in aggregate, are, for the purposes of this part, to be considered a single temporary building.

Combustible materials are to be piled with due regard to the stability of piles, and in no case higher than 20 feet. Driveways between and around combustible storage piles are to be at least 15 feet wide and maintained free from the accumulation of rubbish, equipment, or other articles or materials. Driveways must be so spaced that a maximum grid system unit of 50 feet by 150 feet is produced. Entire storage sites are to be kept free from accumulation of unnecessary combustible materials. Weeds and grass are to be kept down, and a regular procedure must be provided for the periodic cleanup of the entire area. When there is a danger of an underground fire, that land is not to be used for combustible or flammable storage. The method used for piling is, wherever possible, to be solid and in orderly and regular piles. No combustible material is to be stored outdoors within 10 feet of a building or structure.

Indoor storage must not obstruct, or adversely affect the means of exit, and all materials are to be stored, handled, and piled with due regard to their fire characteristics. Noncompatible materials, which may create a fire hazard, are to be segregated by a barrier which has a fire resistance of at least one hour. Material shall be piled to minimize the spread of fire internally, and to permit convenient access for firefighting. Stable piling must be maintained at all times. Aisle space is to be maintained to safely accommodate the widest vehicle that may be used within the building for firefighting purposes. A clearance of at least 36 inches is to be maintained between the top level of the stored material and the sprinkler deflectors. Clearance shall be maintained around lights and heating units to prevent ignition of combustible materials. A clearance of 24 inches is to be maintained around the path of travel of fire doors, unless a barricade is provided, in which case no clearance is needed. Material must not be stored within 36 inches of a fire door opening.

Flammable and Combustible Liquids (1926.152)

When storing or handling flammable and combustible liquids, only approved containers and portable tanks are to be used (see Figure 53). Approved metal safety cans are to be used for the handling and use of flammable liquids in quantities greater than one gallon, except that this does not apply to those flammable liquid materials which are highly viscid (extremely hard to pour); they may be used and handled in original shipping containers. For quantities of one gallon or less, only the original container or approved metal safety cans can be used for storage, use, and handling of flammable liquids.

Flammable or combustible liquids are not to be stored in areas used for exits, stairways, or normally used for the safe passage of people. No more than 25 gallons of flammable or combustible liquids are to be stored in a room outside of an approved storage cabinet. Quantities of flammable and combustible liquid in excess of 25 gallons are to be stored in an acceptable or approved cabinet, and meet the following requirements.

Specially designed and constructed wooden storage cabinets, which meet unique criteria, can be used for flammable and combustible liquids. Approved metal storage cabinets are also acceptable. Cabinets are to be labeled in conspicuous lettering, "Flammable–Keep Fire Away." Not more than 60 gallons of flammable, or 120 gallons of combustible liquids are to be stored in any one storage cabinet. Not more than three such cabinets may be located in a single storage area. Quantities in excess of this must be stored in an inside storage room. Inside storage rooms are to be constructed to meet the required fire-resistive rating for their use. Such construction must comply with the test specifications set forth in Standard Methods of Fire Test of Building Construction and Material, NFPA 251-1969.

Where an automatic extinguishing system is provided, the system must be designed and installed in an approved manner. Openings to other rooms or buildings must be provided with noncombustible liquid-tight raised sills or ramps at least 4 inches in height, or the floor in the storage area must be at least 4 inches below the surrounding floor. Openings are to be provided with approved self-closing fire doors. The room is to be liquid-tight where the walls

Figure 53. Containers for flammable or combustible liquids

join the floor. A permissible alternate to the sill or ramp is an open-grated trench, inside of the room, which drains to a safe location. Where other portions of the building, or other buildings are exposed, windows are to be protected as set forth in the Standard for Fire Doors and Windows, NFPA No. 80-1970, for Class E or F openings. Wood of at least one inch nominal thickness may be used for shelving, racks, dunnage, scuffboards, floor overlay, and similar installations. Materials which will react with water and create a fire hazard are not to be stored in the same room with flammable or combustible liquids.

Electrical wiring and equipment, located in inside storage rooms, are to be approved for Class I, Division 1, Hazardous Locations. Every inside storage room is to be provided with either a gravity or a mechanical exhausting system. Such system shall commence not more than 12 inches above the floor, and be designed to provide for a complete change of air within the room at least 6 times per hour. If a mechanical exhausting system is used, it must be controlled by a switch located outside of the door. The ventilating equipment and any lighting fixtures are to be operated by the same switch. An electric pilot light shall be installed adjacent to the switch, if flammable liquids are dispensed within the room. Where gravity ventilation is provided, the fresh air intake, as well as the exhausting outlet from the room, must be on the exterior of the building in which the room is located.

In every inside storage room there shall be maintained one clear aisle at least 3 feet wide. Containers over 30 gallons capacity are not to be stacked one upon the other. Flammable and combustible liquids in excess of that permitted in inside storage rooms are to be stored outside of buildings.

The quantity of flammable or combustible liquids kept in the vicinity of spraying operations is to be the minimum required for operations, and should ordinarily not exceed a supply for one day, or one shift. Bulk storage of portable containers of flammable or combustible liquids are to be in a separate, constructed building detached from other important buildings, or cut off in a standard manner.

Storage of containers (not more than 60 gallons each) shall not exceed 1,100 gallons in any one pile or area. Piles or groups of containers are to be separated by a 5-foot clearance. Piles or groups of containers are not to be closer than 20 feet to a building. Within 200 feet of each pile of containers, there should be a 12-foot-wide access way to permit the approach of fire control apparatus. The storage area is to be graded in a manner as to divert possible spills away from buildings or other exposures, or is to be surrounded by a curb or earth dike at least 12 inches high. When curbs or dikes are used, provisions must be made for draining off accumulations of ground or rain water, or spills of flammable or combustible liquids. Drains shall terminate at a safe location, and must be accessible to the operation under fire conditions.

Portable tanks are not to be closer than 20 feet from any building. Two or more portable tanks, grouped together, having a combined capacity in excess of 2,200 gallons, are to be separated by a 5-foot-clear area. Individual portable tanks exceeding 1,100 gallons shall be separated by a 5-foot-clear area. Within 200 feet of each portable tank, there must be a 12-foot-wide access way to permit the approach of fire control apparatus.

Portable tanks, not exceeding 660 gallons, are to be provided with emergency venting and other devices, as required by Chapters III and IV of NFPA 30-1969, The Flammable and Combustible Liquids Code.

Portable tanks, in excess of 660 gallons, must have emergency venting and other devices, as required by Chapters II and III of The Flammable and Combustible Liquids Code, NFPA 30-1969. At least one portable fire extinguisher, having a rating of not less than 20-B units, is to be located outside of, but not more than 10 feet from, the door opening into any room used for storage of more than 60 gallons of flammable or combustible liquids, and not less than 25 feet, nor more than 75 feet, from any flammable liquid storage area located outside.

At least one portable fire extinguisher, having a rating of not less than 20-B:C units, is to be provided on all tank trucks or other vehicles used for transporting and/or dispensing flammable or combustible liquids.

Areas in which flammable or combustible liquids are transferred at one time, in quantities greater than 5 gallons from one tank or container to another tank or container, are to be separated from other operations by 25-feet distance, or by construction having a fire resistance of at least one hour. Drainage or other means are to be provided to control spills. Adequate natural or mechanical ventilation must be provided to maintain the concentration of flammable vapor at, or below 10 percent of the lower flammable limit.

Transfer of flammable liquids from one container to another is done only when containers are electrically interconnected (bonded). When flammable or combustible liquids are drawn from, or transferred into vessels, containers, or tanks within a building or outside. Only through a closed piping system, from safety cans, by means of a device drawing through the top, or from a container, or portable tanks, by gravity or pump, through an approved self-closing valve. Transferring by means of air pressure on the container or portable tanks is prohibited. Any dispensing units are to be protected against collision damage. Dispensing devices and nozzles for flammable liquids are to be of an approved type. The dispensing hose must be an approved type, as well as the dispensing nozzle must be an approved automatic-closing type without a latch-open device. Clearly identified, and easily accessible switch(es) are to be provided at a location remote from dispensing devices, in order to shut off the power to all dispensing devices, in the event of an emergency.

Heating equipment, of an approved type, may be installed in the lubrication or service area where there is no dispensing or transferring of flammable liquids, provided the bottom of the heating unit is at least 18 inches above the floor and is protected from physical damage. Heating equipment installed in lubrication or service areas, where flammable liquids are dispensed, is to be of an approved type for garages, and is to be installed at least 8 feet above the floor. No smoking or open flames are permitted in the areas used for fueling, servicing fuel systems for internal combustion engines, or receiving or dispensing flammable or combustible liquids, and conspicuous and legible signs, prohibiting smoking, are to be posted.

The motors of all equipment being fueled are to be shut off during the fueling operation. Each service or fueling area is to be provided with at least one fire extinguisher having a rating of not less than 20-B:C, and located so that an extinguisher will be within 75 feet of each pump, dispenser, underground fill pipe opening, and lubrication or service area. More information regarding specifics on flammable and combustible liquid storage tanks can be found in 29 CFR 1926.152. Also, further information on Fixed Extinguishing Systems (1926.156), Fixed Extinguishing Systems, Gaseous Agent (1926.157), and Fire Detection System (1926.158) in 29 CFR 1926.

FLAGPERSON (1926.201)

The flagperson has a high exposure potential for injury due to moving motor vehicles and equipment The flagperson must maintain constant vigil since traffic flow is always changing and the flagperson must never leave their work post till properly relieved. The flagperson is to use flags, at least 18 inches square, or sign paddles when hand signaling and must be outfitted with a red or orange reflectorized warning vest while flagging. The flagperson should be trained in directing traffic flow and how to place themselves in a safe position as well as have adequate communications with the work crew and other individuals conducting flagging task (see Figure 54).

All required signs and symbols must be visible at all times when work is being done, and removed promptly when the hazard no longer exists. Signaling directions must conform to ANSI D6.1-1971, *Manual on Uniform Traffic Control Devices for Streets and Highways*.

FLOOR AND WALL OPENINGS (1926.501)

See 29 CFR 1926.501.

Figure 54. A flagperson directing traffic flow

FOOD HANDLING (1926.51)

Food handling and food service must meet applicable laws, ordinances, or regulations, as well as sound hygiene principles.

FOOT PROTECTION (1926.96)

Although construction companies do not require safety-toed shoes to be worn by all workers, it would seem that the habit of wearing safety-toed shoes which meet the ANSI Standard Z41.1-1967 should be a mandatory part of the construction personal protective equipment program. The variety and comfort of these shoes today offset the previous complaints, and definitely foster good foot safety on construction sites. The use of safety-toed shoes should be in the same vane as mandatory hardhats and eye protection, when considering the risk reduction of potential foot injuries.

HAND PROTECTION

In regards to hand protection, a good pair of leather gloves is a must for construction. Leather gloves should be worn when climbing, handling materials, being around sharp or jagged materials, when needed as antivibration protection, and when near low voltage electrical circuits (dry leather gloves provide a degree of insulation from electrical shock). If construction workers do not remove their rings prior to work, the wearing of gloves helps prevent the rings from catching or snagging on objects which can cause cuts or contusion of the fingers. Construction workers' hands and fingers are in constant use and thus, a high level of exposure and risk of injury exist. Hand protection makes a lot of sense.

At times, special hand protection may be needed when exposure to chemical or high voltage electricity exists. This is when the proper selection of hand protection is so vital. The use of commercially generated glove selection guides, provided by manufacturers, need to be consulted to assure that workers are using the right glove for the job.

With the high number of hand and finger injuries, it is well worth requiring the use of hand protection in the form of gloves.

HAND AND POWER TOOLS/GUARDING (1926.300)

All hand and power tools, and similar equipment, whether furnished by the employer or the employee, are to be maintained in a safe condition. When power-operated tools are designed to accommodate guards, they must be equipped with such guards when in use. Any belts, gears, shafts, pulleys, sprockets, spindles, drums, fly wheels, chains, or other reciprocating, rotating, or moving parts of equipment are to be guarded, if such parts are exposed to contact by workers, or otherwise create a hazard. Guarding must meet the requirements as set forth in American National Standards Institute, B15.1-1953 (R1958), *Safety Code for Mechanical Power-Transmission Apparatus.*

One or more methods of machine guarding must be provided to protect the operator and other employees in the machine area from hazards such as those created by point of operation, in going nip points, rotating parts, flying chips, and sparks. Examples of guarding methods are: barrier guards, two-hand tripping devices, electronic safety devices, etc. Point of operation is the area on a machine where work is actually performed upon the material being processed. The point of operation of machines where operation exposes a worker to injury, is to be guarded. The guarding is to be in conformity with any appropriate standards therefore, or in the absence of applicable specific standards, is so designed and constructed as to prevent the operator from having any part of his body in the danger zone during the operating cycle.

Special handtools for placing and removing material are to be such as to permit easy handling of material without the operator placing a hand in the danger zone. Such tools are not to be used in lieu of other guarding required by this section, but can only be used to supplement protection provided. The following are some of the machines which usually require point of operation guarding: guillotine cutters, shears, alligator shears, power presses, milling machines, power saws, jointers, portable power tools, forming rolls, and calenders. Machines designed for a fixed location shall be securely anchored to prevent walking or moving.

When the periphery of the blades of a fan is less than 7 feet (2.128 m) above the floor or working level, the blades must be guarded. The guard must have openings no larger than 1/2 inch (1.27 cm).

Employees using hand and power tools, and exposed to the hazard of falling, flying, abrasive, and splashing objects, or exposed to harmful dusts, fumes, mists, vapors, or gases are to be provided with the particular personal protective equipment necessary to protect them from the hazard. All personal protective equipment shall meet the requirements and be maintained according to Subparts D and E of 29 CFR 1926.

All hand-held powered platen sanders, grinders with wheels 2-inch diameter or less, routers, planers, laminate trimmers, nibblers, shears, scroll saws, and jigsaws with blade shanks one-fourth of an inch wide or less, are to be equipped with only a positive "on-off" control. All hand-held powered drills, tappers, fastener drivers, horizontal, vertical, and angle grinders with wheels greater than 2 inches in diameter, disc sanders, belt sanders, reciprocating saws, saber saws, and other similar operating powered tools, are to be equipped with a momentary contact "on-off" control, and may have a lock-on control, provided that turnoff can be accomplished by a single motion of the same finger, or fingers that turn it on. All other hand-held powered tools, such as circular saws, chain saws, and percussion tools without positive accessory holding means, are to be equipped with a constant pressure switch that will shut off the power when the pressure is released. This paragraph does not apply to concrete vibrators, concrete breakers, powered tampers, jack hammers, rock drills, and similar hand-operated power tools.

HAND TOOLS (1926.301)

Tools are such a common part of construction work that it is difficult to remember that they may pose hazards. In the process of removing or avoiding the hazards, workers must learn to recognize the hazards associated with the different types of tools, and the safety precautions necessary to prevent injury from those hazards. Therefore, in an effort to minimize accidents resulting from the use of hand tools, certain precautions need to be taken, such as:

1. Do not use broken, defective, burned, or mushroomed tools. Report defective tools to your supervisor and turn tools in for replacement.

2. Always use the proper tool and equipment for any task you may be assigned to do. For example: do not use a wrench as a hammer, or a screwdriver as a chisel.

3. Do not leave tools on scaffolds, ladders, or any overhead working surfaces. Racks, bins, hooks, or other suitable storage space must be provided and arranged to permit convenient arrangement of tools.

4. Do not strike two hardened steel surfaces together (i.e., two hammers, or a hammer and hardened steel shafts, bearings, etc.).

5. Do not throw tools from one location to another, from one worker to another, or drop them to lower levels; this is prohibited. When necessary to pass tools or material under the above conditions, suitable containers and/or ropes must be used.

6. Wooden tool handles must be sound, smooth, and in good condition, and securely fastened to the tool (see Figure 55).

7. Sharp-edged or pointed tools should never be carried in employee's pockets.

8. Only non-sparking tools shall be used in locations where sources of ignition may cause a fire or explosion.

9. Tools requiring heat treating should be tempered, formed, dressed, and sharpened by workmen experienced in these operations. Wrenches, including adjustable, pipe, end, and socket wrenches are not to be used when jaws are sprung to the point that slippage occurs.

10. Any defective tool should be removed from service and tagged indicating it is not to be used.

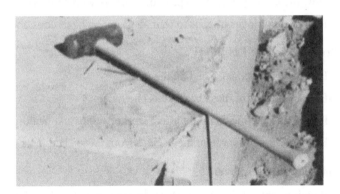

Figure 55. Wooden handled sledge hammer in good repair

HAZARD COMMUNICATIONS (1926.59)

OHSA has established regulations for the general industry, and the construction industry, called the Hazard Communication Standard (29 CFR 1926.59 & 29 CFR 1910.1200). These standards require that manufacturers of hazardous chemicals inform employers about the hazards of those chemicals. Also, it requires employers to inform employees of the identities, properties, characteristics, and hazards of chemicals they use, and the protective measures they can take to prevent adverse effects. The standard covers both physical hazards (e.g., flammability), and health hazards (e.g., lung damage, cancer). Knowledge acquired under the Hazard Communication Standard will help employers provide safer work places for workers, establish proper work practices, and help prevent chemical-related illnesses and injuries. Employers are required to do the following:

*   The employer must develop a written hazard communication program.

*   The employer must provide specific information and training to workers.

- All employers on a multiple employer site must provide information to each other so that all employees can be protected.

- The owner must provide information to contractors about hazardous materials on the job site.

The specific requirements for each of the four main provisions are summarized as follows.

Written Hazard Communication Program

1. List of hazardous chemicals on the jobsite.

2. The method the employer will use to inform employees of the hazards associated with non-routine tasks involving hazardous chemicals.

3. How the employer plans to provide employees of other companies on the jobsite with the material safety data sheets (MSDSs), such as making them available at a central location.

4. The method the employer will use to inform employees of other companies on the jobsite about their labeling system.

5. How the employer will inform workers about their labeling system.

6. How the employer plans to provide workers with MSDSs.

7. How the employer intends to train workers on hazardous chemicals.

Information Provided by the Employer

1. Hazardous chemicals used on the job.

2. How to recognize these hazardous chemicals.

3. How those chemicals might affect worker safety and health.

4. How workers can protect themselves from those chemicals (see Figure 56).

Training Provided by the Employer

1. Requirements of the OSHA Hazard Communication Standard.

2. Operations at the worksite where hazardous chemicals are present.

3. The location and availability of the

    * Written Hazard Communication Program.

    * List of all hazardous chemicals.

    * MSDSs for all hazardous chemicals used on the jobsite.

4. Methods and observations workers can use to detect the presence or release of hazardous chemicals in your work area {e.g., labels, color, form (solid, liquid, gas), and order}.

5. The physical and health hazards workers may be exposed to from the hazardous chemicals on the job.

6. Methods of protecting oneself, such as work practices, personal protective equipment, and emergency procedure.

7. Details of the hazardous communication program used by the employer.

8. Explanation of how workers can obtain and use hazard information.

Figure 56. Many different chemicals are found on construction sites

Multiple Employer Sites

1. All employers on a multiple employer site must supply information to each other, so all employees will be protected.

2. The hazard communication program must specify how an employer will provide other employers with a copy of the MSDSs, or make it available at a central location in the work place, for each hazardous chemical the other employer(s)' employees may be exposed to while working.

3. The employers must provide the procedures for informing other employer(s) of any precautionary measures that need to be taken to protect employees during the worksite's normal working operating conditions, and of any foreseeable emergencies.

4. An employer must provide the mechanism to inform other employer(s) of his or her labeling system.

Owner's/Contractor's Responsibilities

In summary, employers are responsible to develop a hazard communication program and provide information to employees and other employer's employees and provide training to employees. All workers, as well as other employees on multiple employer worksites, must be provided with information regarding any hazardous chemicals to which workers might be exposed to at the employers' place of work.

Site owners, such as owners of chemical companies or paper pulp mills, must inform the contractor about the hazardous chemicals used at their plant which could result in exposure to the construction workers who will be working at their worksite. All employers in construction, general industry, etc. must comply with the hazardous communication regulations.

HAZARDOUS WASTE OPERATIONS (1926.65)

Hazard waste operations include the following operations, unless the employer can demonstrate that the operation does not involve employee exposure, or the reasonable possibility for employee exposure to safety or health hazards.

1.   Clean-up operations required by a governmental body, whether Federal, state, local, or others involving hazardous substances, which are conducted at uncontrolled hazardous waste sites (including, but not limited to, the EPA's National Priority Site List (NPL), state priority site lists, sites recommended for the EPA NPL, and initial investigations of government identified sites which are conducted before the presence or absence of hazardous substances has been ascertained).

2.   Corrective actions involving clean-up operations at sites covered by the Resource Conservation and Recovery Act of 1976 (RCRA) as amended in (42 U.S.C. 6901 *et seq.*).

3.   Voluntary clean-up operations at sites recognized by Federal, state, local, or other governmental bodies as uncontrolled hazardous waste sites.

4.   Operations involving hazardous wastes that are conducted at treatment, storage, and disposal (TSD) facilities regulated by 40 CFR parts 264 and 265 pursuant to RCRA; or by agencies under agreement with U.S.E.P.A. to implement RCRA regulations.

5.   Emergency response operations for releases of, or substantial threats of releases of, hazardous substances without regard to the location of the hazard.

Employers must comply with paragraph (p) of 1926.65 unless conducting emergency response operations, then employers must comply with paragraph (q) of 1926.65. Hazardous waste operations pose not only chemical hazards, but most of the hazards which are usually faced by those working in the construction industry. Hazardous waste sites are really construction sites with the added hazards involved with the handling, removal, and disposal of hazardous chemicals.

Written Safety and Health Program

As part of paragraph (p), employers must develop and implement a written safety and health program for their employees involved in hazardous waste operations. The program is to be designed to identify, evaluate, and control safety and health hazards, and provide for emergency response for hazardous waste operations. The written safety and health program shall contain the following:

1.   An organizational structure.

2.   A comprehensive workplan.

3.   A site-specific safety and health plan which need not repeat the employer's standard operating procedures.

4.   The safety and health training program.

5.   The medical surveillance program.

6.   The employer's standard operating procedures for safety and health.

7.   Any necessary interface between the general program and site specific activities.

An employer, who retains contractor or subcontractor services for work in hazardous waste operations, must inform those contractors, subcontractors, or their representatives of the site emergency response procedures, and any potential fire, explosion, health, safety, or other hazards of the hazardous waste operation that have been identified by the employer, including those identified in the employer's information program. The written safety and health program is to be made available to: any contractor, subcontractor, or their representative, who will be involved with the hazardous waste operation; to employees; to employee designated representatives; to OSHA personnel; and to personnel of other federal, state, or local agencies with regulatory authority over the site.

The organizational structure part of the program must establish the specific chain of command and specify the overall responsibilities of the supervisors and employees. It must include, at a minimum, the following elements: a general supervisor who has the responsibility and authority to direct all hazardous waste operations; a site safety and health supervisor who has the responsibility and authority to develop and implement the site safety and health plan and verify compliance; and all other personnel needed for the hazardous waste site operations and emergency response. They must be informed of their general functions, responsibilities, and the lines of authority, responsibility, and communication. The organizational structure is to be reviewed and updated, as necessary, to reflect the current status of the waste site operations.

The comprehensive workplan part of the program addresses the tasks and objectives of the site operations, and the logistics and resources required to reach those tasks and objectives. The comprehensive workplan addresses anticipated clean-up activities, as well as normal operating procedures, which need not repeat the employer's procedures available elsewhere. The comprehensive workplan defines work tasks and objectives and identifies the methods for accomplishing those tasks and objectives; establishes personnel requirements for implementing the plan; provides for the implementation of the training required; provides for the implementation of the required informational programs; and provides for the implementation of the medical surveillance program.

Site Safety and Health Plan

The site safety and health plan, which must be kept on site, addresses the safety and health hazards of each phase of the site operation, and includes the requirements and procedures for employee protection, as a minimum, and addresses the following:

1. A safety and health risk or hazard analysis for each site task and operation found in the workplan.

2. Employee training assignments to assure compliance.

3. Personal protective equipment to be used by employees for each of the site tasks and operations being conducted, as required by the personal protective equipment program.

4. Medical surveillance requirements in accordance with the program.

5. Frequency and types of air monitoring, personnel monitoring, and environmental sampling techniques and instrumentation to be used, including methods of maintenance and calibration of monitoring and sampling equipment to be used.

6. Site control measures used in accordance with the site control program.

7. Decontamination procedures.

8. An emergency response plan for safe and effective responses to emergencies, including the necessary PPE and other equipment.

9. Confined space entry procedures.

10. A spill containment program.

The site specific safety and health plan must provide for pre-entry briefings to be held prior to initiating any site activity, and at such other times as necessary to ensure that employees are apprised of the site safety and health plan, and that this plan is being followed. The information and data obtained from site characterization and analysis work are to be used to prepare and update the site safety and health plan.

Inspections are to be conducted by the site safety and health supervisor or, in the absence of that individual, another individual who is knowledgeable in occupational safety and health who acts on behalf of the employer, as necessary, to determine the effectiveness of the site safety and health plan. Any deficiencies in the effectiveness of the site safety and health plan shall be corrected by the employer.

Site Evaluation

Hazardous waste sites are to be evaluated to identify specific site hazards and to determine the appropriate safety and health control procedures needed to protect employees from the identified hazards.

A preliminary evaluation of a site's characteristics is to be performed prior to site entry, by a qualified person, in order to aid in the selection of appropriate employee protection methods prior to site entry. Immediately after initial site entry, a more detailed evaluation of the site's specific characteristics is to be performed by a qualified person in order to further identify existing site hazards, and to further aid in the selection of the appropriate engineering controls and personal protective equipment for the tasks to be performed.

All suspected conditions that may pose inhalation or skin absorption hazards that are immediately dangerous to life or health (IDLH), or other conditions that may cause death or serious harm, must be identified during the preliminary survey and evaluated during the detailed survey. Examples of such hazards include, but are not limited to, confined space entry, potentially explosive or flammable situations, visible vapor clouds, or areas where biological indicators, such as dead animals or vegetation, are located.

The following information, to the extent available, shall be obtained by the employer prior to allowing employees to enter a site:

1. Location and approximate size of the site.

2. Description of the response activity and/or the job task to be performed.

3. Duration of the planned employee activity.

4. Site topography and accessibility by air and roads.

5. Safety and health hazards expected at the site.

6. Pathways for hazardous substance dispersion.

7. Present status and capabilities of emergency response teams that would provide assistance to hazardous waste clean-up site employees at the time of an emergency.

8. Hazardous substances and health hazards involved or expected at the site, and their chemical and physical properties.

Personal protective equipment (PPE) is to be provided and used during initial site entry. Based upon the results of the preliminary site evaluation, an ensemble of PPE is to be selected and used during the initial site entry, which will provide protection to a level of exposure below permissible exposure limits, and published exposure levels for known or suspected hazardous substances and health hazards, and which will provide protection against other known and suspected hazards identified during the preliminary site evaluation. If there is no permissible exposure limit or published exposure level, the employer may use other published studies and information as a guide to appropriate personal protective equipment. If positive-pressure, self-contained breathing apparatus is not used as part of the entry ensemble, and if respiratory protection is warranted by the potential hazards identified during the preliminary site evaluation, an escape self-contained breathing apparatus of at least five minute's duration is to be carried by employees during the initial site entry. If the preliminary site evaluation does not produce sufficient information to identify the hazards or suspected hazards of the site, an ensemble providing protection equivalent to Level B PPE is to be provided as minimum protection, and direct reading instruments must be used, as appropriate, for identifying IDLH conditions. Once the hazards of the site have been identified, the appropriate PPE is selected and used.

When the site evaluation produces information that shows the potential for ionizing radiation or IDLH conditions, or when the site information is not sufficient to reasonably eliminate these possible conditions, then the following methods are to be used: monitoring with direct reading instruments for hazardous levels of ionizing radiation; monitoring the air with appropriate direct reading test equipment (i.e., combustible gas meters, detector tubes) for IDLH and other conditions that may cause death or serious harm (combustible or explosive atmospheres, oxygen deficiency, toxic substances); monitoring by visually observing for signs of actual or potential IDLH or other dangerous conditions; and monitoring by an ongoing air monitoring program. All these methods are to be implemented after site characterization has determined the site is safe for the start-up of operations.

Once the presence and concentrations of specific hazardous substances and health hazards have been established, the risks associated with these substances shall be identified. Employees who will be working on the site are to be informed of any risks that have been identified, and receive the required training, including hazard communications. The risks to consider include, but are not limited to, exposures exceeding the permissible exposure limits and published exposure levels, IDLH concentrations, potential skin absorption and irritation sources, potential eye irritation sources, explosion sensitivity and flammability ranges, and oxygen deficiency.

Any information that is available to the employer, concerning the chemical, physical, and toxicological properties of each substance known, or expected to be present on the jobsite, and is relevant to the duties an employee who is expected to perform at that jobsite, shall be made available to the affected employees prior to the commencement of their work activities. The employer may utilize information developed for the hazard communication standard for this purpose.

Site Control

Before clean-up work begins, appropriate site control procedures are to be implemented to control employee exposure to hazardous substances. Therefore, a site control program must be developed, for the protection of the employees, during the planning stages of a hazardous waste clean-up operation, and it is to be a part of the employer's site safety and health program. The program shall be modified, as necessary, as new information becomes available.

The site control program shall, as a minimum, include: a site map, site work zones, the use of a "buddy system," site communications (including alerting means for emergencies), standard operating procedures or safe work practices, and identification of the nearest medical assistance.

All employees working on site, such as, but not limited to, equipment operators, general laborers, and others exposed to hazardous substances, health hazards, or safety hazards, and their supervisors and management responsible for the site, shall receive training before they are permitted to engage in hazardous waste operations that could expose them to hazardous substances, safety, or health hazards.

Workers are to be trained to the level required by their job function and responsibility. The training must thoroughly cover the following:

1.   Names of personnel and alternates responsible for site safety and health.

2.   Safety, health, and other hazards present on the site.

3.   Use of personal protective equipment.

4.   Work practices by which the employee can minimize risks from hazards.

5.   Safe use of engineering controls and equipment on the site.

6.   Medical surveillance requirements, including recognition of symptoms and signs which might indicate overexposure to hazards.

7.   The contents of the paragraphs of the site safety and health plan.

Training

General site workers (such as equipment operators, general laborers, and supervisory personnel) engaged in hazardous substance removal, or other activities which expose or potentially expose workers to hazardous substances and health hazards, must receive a minimum of 40 hours of instruction off the site, and a minimum of three days actual field experience under the direct supervision of a trained, experienced supervisor. Workers who are only on the site occasionally for a specific limited task (such as, but not limited to, groundwater monitoring, land surveying, or geo-physical surveying), and who are unlikely to be exposed over permissible and published exposure limits are to receive a minimum of 24 hours of instruction off the site, and the minimum of one day actual field experience under the direct supervision of a trained, experienced supervisor. Workers who are regularly on the site, and work in areas which have been monitored, and these areas have been fully characterized and indicate that exposures are under the permissible and published exposure limits and do not require the use of respirators, and that the characterization indicates that there are no health hazards, or the possibility of an emergency developing, must receive a minimum of 24 hours of instruction off the site and the minimum of one day actual field experience under the direct supervision of a trained, experienced supervisor. Workers with 24 hours of training, who are covered by this paragraph, who become general site workers, or who are required to wear respirators, must have the additional 16 hours and two days of training (see Figure 57).

On-site management and supervisors who are directly responsible for, or who supervise employees engaged in, hazardous waste operations must receive 40 hours initial training, and three days of supervised field experience. The training may be reduced to 24 hours and one day, and/or at least eight additional hours of specialized training, at the time of job assignment, on such topics as, but not limited to, the employer's safety and health program, the associated employee training program, the personal protective equipment program, the spill containment program, and the health hazard monitoring procedures and techniques.

Trainers are to be qualified to instruct employees in the subject matter that is being presented in training. Such trainers must have satisfactorily completed a training program for teaching the subjects they are expected to teach, or they must have the academic credentials and instructional experience necessary for teaching the subjects. Instructors are to demonstrate competent instructional skills and knowledge of the applicable subject matter.

Employees and supervisors who have received and successfully completed the training and have the field experience are to be certified by their instructor or the head instructor and trained supervisor as having successfully completed the necessary training. A written certificate is given to each person so certified. Any person who has not been so certified, or who

Figure 57. Trained workers doing hazardous waste work

does not meet the requirements, is prohibited from engaging in hazardous waste operations.

Employees who respond to hazardous emergency situations at hazardous waste clean-up sites, and may be exposed to hazardous substances, are to be trained how to respond to such expected emergencies.

Employees, managers, and supervisors, specified, are to receive eight hours of refresher training, annually, on specified items. Critiques may be used of incidents that have occurred in the past year; these can serve as training examples for related work and other relevant topics.

Employers, who can show by documentation or certification that an employee's work experience, and/or training, has resulted in training equivalent to that training required, are not required to provide the initial training requirements to these employees; the employer must provide a copy of the certification or documentation to the employee upon request. However, certified employees, or employees who have equivalent training, and who are new to a site, must receive the appropriate site specific training before site entry, and must have the appropriate supervised field experience at the new site. Equivalent training includes any academic training or the training that existing employees might have already received from actual hazardous waste site work experience.

Medical Surveillance

Employers engaged in hazardous waste operations must institute a medical surveillance program which includes employees who are, or may be exposed to hazardous substances or health hazards at, or above the permissible exposure limits, or if there is no permissible exposure limit above the published exposure levels for these substances, without regard to the use of respirators, for 30 days or more a year. This also includes all employees who wear a respirator for 30 days or more a year, or as required by 1926.103; all employees who are injured, become ill, or develop signs or symptoms due to possible overexposure which involves hazardous substances or health hazards from an emergency response or hazardous waste operation; and members of HAZMAT teams.

Medical examinations and consultations are to be made available by the employer to each employee, as follows: prior to an assignment; at least once every twelve months for each employee covered (unless the attending physician believes a longer interval, not greater than biennially, is appropriate); at the termination of employment; when an employee is reassigned to an area and the employee would not be covered if the employee has not had an examination within the last six months; as soon as possible after notification by an employee, that the employee has developed signs or symptoms indicating possible overexposure to hazardous substances or health hazards; when an employee has been injured, or exposed above the permissible exposure limits, or published exposure levels in an emergency situation; and at more frequent times, if the examining physician determines that an increased frequency of examination is medically necessary.

All employees are to be provided medical examinations, including: those who may have been injured; who received a health impairment; those who developed signs or symptoms which may have resulted from exposure to hazardous substances while working at an emergency incident; those who may have been exposed, during an emergency incident, to hazardous substances at concentrations above the permissible exposure limits or published exposure levels without the necessary personal protective equipment being used; as soon as possible following the emergency incident or development of signs or symptoms and at additional times, if the examining physician determines that follow-up examinations or consultations are medically necessary.

Medical examinations must include a medical and work history, or updated history if one is in the employee's file, and should have a special emphasis on the employee's symptoms related to the handling of hazardous substances and health hazards, and the employee's fitness for duty, including the ability to wear any required PPE under conditions (i.e., temperature extremes) that may be expected at the worksite. The content of the medical examinations or consultations made available to employees is to be determined by the attending physician. The guidelines in the *Occupational Safety and Health Guidance Manual for Hazardous Waste Site Activities* should be consulted.

All medical examinations and procedures are to be performed by, or under the supervision of a licensed physician, preferably one knowledgeable in occupational medicine, and are to be provided without cost to the employee, without loss of pay, and at a reasonable time and place. The employer shall provide one copy of hazardous waste standards, and its appendices, to the attending physician. In addition, the following must be supplied for each employee: a description of the employee's duties as they relate to the employee's exposures; the employee's exposure levels, or anticipated exposure levels; a description of any personal protective equipment used, or to be used; information from the employee's previous medical examinations which is not readily available to the examining physician; and information required by 1926.103.

The employer must obtain and furnish the employee with a copy of a written opinion from the attending physician which contains the following:

1. The physician's opinion as to whether the employee has any detected medical conditions which would place the employee at an increased health risk of material impairment when working at hazardous waste operations, when working in emergency response situations, or when using a respirator.

2. The physician's recommended limitations upon the employee's assigned work.

3. The results of the medical examination and tests, if requested by the employee.

4. A statement that the employee has been informed by the physician of the results of the medical examination and any medical conditions which require further examination or treatment.

The written opinion obtained by the employer is not to reveal specific findings or diagnoses unrelated to occupational exposures. An accurate record of the medical surveillance is to be retained for the length of employment, plus 30 years. The required record must include at least the following information:

1.   The name and social security number of the employee.

2.   The physician's written opinions, recommended limitations, and results of examinations and tests.

3.   Any employee medical complaints related to exposure to hazardous substances.

4.   A copy of the information provided to the examining physician by the employer, with the exception of the standard and its appendices.

Hazard Controls

Engineering controls, work practices, personal protective equipment, or a combination of these, are to be implemented to protect employees from exposure to hazardous substances, and safety and health hazards. Engineering controls, when feasible, should include the use of pressurized cabs or control booths on equipment, and/or the use of remote-operated material handling equipment. Also, work practices that should be followed, when feasible, are those of removing all non-essential employees from potential exposure during the opening of drums and the wetting down dusty operations; locate these employees upwind of possible hazards.

Whenever engineering controls and work practices are not feasible, or not required, any reasonable combination of engineering controls, work practices, and PPE are to be used to reduce and maintain employee exposures to, or below the permissible exposure or dose limits for substances. The employer is not to use a schedule of employee rotation as a means of compliance with permissible exposure or dose limits, except when there is no other feasible way of complying with the airborne or dermal dose limits for ionizing radiation. The employer may use the published literature and MSDS as a guide in making a determination as to what level of protection the employer believes is appropriate for hazardous substances and health hazards, when there are no permissible or published exposure limits.

Personal Protective Equipment

Personal protective equipment (PPE) must be selected and used which will protect employees from the hazards and potential hazards that are identified during the site characterization and analysis. Personal protective equipment selection is to be based on an evaluation of the performance characteristics of the PPE relative to the requirements and limitations of the site, the task-specific conditions and duration, and the hazards and potential hazards identified at the site.

Positive pressure, self-contained breathing apparatus, or positive pressure, air-line respirators which are equipped with an escape air supply, are to be used when the chemical exposure levels present at the site create a substantial possibility of immediate death, immediate serious illness or injury, or impair the ability to escape.

Totally encapsulating chemical protective suits are to be used in conditions where skin absorption of a hazardous substance may result in a substantial possibility of immediate death, immediate serious illness or injury, or impair the ability to escape.

The level of protection provided by PPE selection is to be increased when additional information on site conditions indicates that increased protection is necessary in order to reduce employees exposure to below the permissible exposure limits, the published exposure levels for hazardous substances, and health hazards. The level of employee protection provided may be decreased when additional information or site conditions show that decreased protection will not result in hazardous exposure to employees.

A written personal protective equipment program must be established which is part of the employer's safety and health program and also a part of the site-specific safety and health plan. The PPE program must address the elements listed below:

1. PPE selection based upon site hazards.

2. PPE use and limitations of the equipment.

3. Work mission duration.

4. PPE maintenance and storage.

5. PPE decontamination and disposal.

6. PPE training and proper fitting.

7. PPE donning and doffing procedures.

8. PPE inspection procedures prior to, during, and after use.

9. Evaluation of the effectiveness of the PPE program.

10. Limitations during temperature extremes, heat stress, and other appropriate medical considerations.

Monitoring

If there are no permissible exposure limits listed for hazardous substances being used at the jobsite, monitoring is to be performed in order to assure the proper selection of engineering controls, work practices, and personal protective equipment so that employees are not exposed to levels which exceed the permissible exposure limits, or published exposure levels. Air monitoring is used to identify and quantify airborne levels of hazardous substances, and safety and health hazards, in order to determine the appropriate level of employee protection needed on site.

Upon initial entry, representative air monitoring is to be conducted to identify any IDLH condition, exposure over permissible exposure limits or published exposure levels, exposure over a radioactive material's dose limits, or other dangerous conditions such as the presence of flammable atmospheres or oxygen-deficient environments.

Periodic monitoring is to be conducted when the possibility of an IDLH condition or flammable atmosphere has developed, or when there is an indication that exposures may have risen over the permissible exposure limits or published exposure levels after the initial monitoring. These conditions may be: work beginning on a different portion of the site; contaminants, other than those previously identified, being handled; a different type of operation being initiated (e.g., drum opening, as opposed to exploratory well drilling); employees handling leaking drums or containers; or employees working in areas with obvious liquid contamination (e.g., a spill or lagoon).

Handling and Transporting Hazardous Materials

Hazardous substances and contaminated soils, liquids, and other residues are to be handled, transported, labeled, and disposed following accepted guidelines. Drums and containers used during the clean-up must meet the appropriate DOT, OSHA, and EPA regulations for the wastes that they contain. When practical, drums and containers are to be inspected and their integrity assured prior to being moved. Drums or containers that cannot be inspected before being moved, because of storage conditions (i.e., buried beneath the earth, stacked behind other drums, stacked several tiers high in a pile, etc.), are to be moved to an accessible location and inspected prior to further handling. Unlabeled drums and containers are considered to contain hazardous substances and handled accordingly until the contents are positively identified and labeled.

Site operations should be organized to minimize the amount of drum or container movement. Prior to movement of drums or containers, all employees exposed to the transfer operation are to be warned of the potential hazards associated with the contents of the drums or containers. The U.S. Department of Transportation's specified salvage drums or containers, and suitable quantities of proper absorbent, are to be kept available and used in areas where spills, leaks, or ruptures may occur. Where major spills may occur, a spill containment program, which is part of the employer's safety and health program, is to be implemented; it is to contain and isolate the entire volume of the hazardous substance being transferred. Drums and containers that cannot be moved without rupture, leakage, or spillage are to be emptied into a sound container, using a device classified for the material being transferred.

Where drums or containers are being opened and an airline respirator system is used, connections to the source of air supply are to be protected from contamination and the entire system must be protected from physical damage. Employees not actually involved in the opening of drums or containers are to be kept a safe distance. If employees must work near or adjacent to drums or containers being opened, a suitable shield, that does not interfere with the work operation, is to be placed between the employee and the drums or containers being opened; this will protect the employee in case of accidental explosion. Controls for drum or container opening equipment, monitoring equipment, and fire suppression equipment are to be located behind the explosion-resistant barrier.

Drums and containers are to be opened in such a manner that excess interior pressure will be safely relieved. If pressure cannot be relieved from a remote location, appropriate shielding must be placed between the employee and the drums or containers in order to reduce the risk of employee injury. Employees must not stand upon, or work from drums or containers.

Material handling equipment used to transfer drums and containers must be selected, positioned, and operated to minimize sources of ignition, related to the equipment, from igniting vapors released from ruptured drums or containers. Drums and containers containing radioactive wastes are not to be handled until such time as their hazard to employees is properly assessed.

As a minimum, the following special precautions are to be taken when drums and containers containing, or suspected of containing shock-sensitive wastes are handled. All non-essential employees are to be evacuated from the area of transfer. Material handling equipment must be provided with explosive containment devices, or protective shields, to protect equipment operators from exploding containers. An employee alarm system, capable of being perceived above surrounding light and noise conditions, is to be used to signal the commencement and completion of explosive waste handling activities. Continuous communications (i.e., portable radios, hand signals, telephones, as appropriate) are to be maintained between the employee-in-charge of the immediate handling area, both the site safety and health supervisors, and the command post, until such time as the handling operation is completed. Communication equipment, or methods that could cause shock sensitive materials to explode, shall not be used. Drums and containers under pressure, as evidenced by bulging or swelling, are not to be moved until such time as the cause for excess pressure is determined, and appropriate containment procedures have been implemented to protect employees from explosive relief of the drum. Drums and containers containing packaged laboratory wastes are to be considered to contain shock-sensitive or explosive materials until they have been characterized.

Lab Packs

Lab packs are to be opened only when necessary, and then only by an individual knowledgeable in the inspection, classification, and segregation of the containers within the pack, according to the hazards of the wastes. If crystalline material is noted on any container, the contents are to be handled as a shock-sensitive waste until the contents are identified. Sampling of containers and drums is to be done in accordance with a sampling procedure, which is part of the site safety and health plan.

Drum or Container Staging

Drums and containers are to be identified and classified prior to packaging for shipment. Drum or container staging areas are to be kept to the minimum number necessary to identify and classify materials safely and to prepare them for transport. Staging areas are to be provided with adequate access and egress routes.

Bulking of hazardous wastes is permitted only after a thorough characterization of the materials has been completed. Tanks and vaults containing hazardous substances are to be handled in a manner similar to that for drums and containers, taking into consideration the size of the tank or vault. Appropriate tank or vault entry procedures, as described in the employer's safety and health plan, are to be followed whenever employees must enter a tank or vault.

Decontamination

Procedures for all phases of decontamination are to be developed and implemented. A decontamination procedure is to be developed, communicated to employees, and implemented before any employees or equipment may enter areas on site where the potential for exposure to hazardous substances exists. All employees leaving a contaminated area are to be appropriately decontaminated and all contaminated clothing and equipment leaving a contaminated area must be appropriately disposed of, or decontaminated. Decontamination procedures are to be monitored by the site safety and health supervisor to determine their effectiveness.

Decontamination must be performed in geographical areas that will minimize exposure of uncontaminated employees or equipment to contaminated employees or equipment. All equipment and solvents used for decontamination are to be decontaminated or disposed of properly.

Protective clothing and equipment are to be decontaminated, cleaned, laundered, maintained, or replaced, as needed, to maintain their effectiveness. Employees, whose non-impermeable clothing becomes wetted with hazardous substances, must immediately remove that clothing and proceed to the shower. The clothing is to be disposed of, or decontaminated before it is removed from the work zone. Unauthorized employees do not remove protective clothing or equipment from change rooms. Commercial laundries or cleaning establishments that decontaminate the protective clothing or equipment are to be informed of the potentially harmful effects of exposures to hazardous substances.

Emergency Response Plan

To handle anticipated emergencies, and prior to the commencement of hazardous waste operations, an emergency response plan is to be developed and implemented by all employers. The plan must be in writing and be available for inspection and copying by employees, their representatives, OSHA personnel, and other governmental agencies with relevant responsibilities. Employers who will evacuate their employees from the danger area when an emergency occurs, and who do not permit any of their employees to assist in handling the emergency, if they provide an emergency action plan complying with 1926.35 are exempt from the other parts of this paragraph. The employer who develops an emergency response plan for emergencies must address, as a minimum, the following:

1. Pre-emergency planning.

2. Personnel roles, lines of authority, and communication.

3. Emergency recognition and prevention.

4. Safe distances and places of refuge.

5. Site security and control.

6. Evacuation routes and procedures.

7. Decontamination procedures which are not covered by the site safety and health plan.

8. Emergency medical treatment and first aid.

9. Emergency alerting and response procedures.

10. Critique of response and follow-up.

11. PPE and emergency equipment.

Emergency response plans must include site topography, layout, and prevailing weather conditions, and procedures for reporting incidents to local, state, and federal governmental agencies.

The emergency response plan is to be a separate section of the site safety and health plan, and shall be compatible and integrated with the disaster, fire, and/or emergency response plans of local, state, and federal agencies. The emergency response plan is to be rehearsed regularly as part of the overall training program for site operations and is to be reviewed periodically and as necessary, be amended to keep it current with new or changing site conditions or information. An employee alarm system is to be installed in accordance with 29 CFR 1926.159.

Sanitation

All toilets, potable water, showers, washing facilities, change rooms, temporary sleeping quarters, and food service facilities must follow existing construction regulations. When hazardous waste clean-up or removal operations commence on a site, and the duration of the work will require six months or greater time to complete, the employer must provide showers and change rooms for all employees who will be exposed to hazardous substances and health hazards. Showers and change rooms are to be located in areas where exposures are below the permissible exposure limits and published exposure levels. If this cannot be accomplished, then a ventilation system must be provided that will supply air that is below the permissible exposure limits and published exposure levels. Employers need to assure that employees shower at the end of their work shift and when leaving the hazardous waste site.

New Technologies

The employer must develop and implement procedures for the introduction of effective new technologies and equipment developed for the improved protection of employees working with hazardous waste clean-up operations. The same is to be implemented as part of the site safety and health program to assure that employee protection is being maintained. New technologies, equipment, or control measures available to the industry, such as the use of foams, absorbents, neutralizers, or other means used to suppress the level of air contaminates while excavating the site or for spill control, are to be evaluated by employers or their representatives. Such an evaluation is to be done to determine the effectiveness of the new methods, materials, or equipment before implementing their use on a large scale. Information and data from manufacturers or suppliers may be used as part of the employer's evaluation effort. Such evaluations shall be made available to OSHA upon request.

RCRA

Resource Conservation and Recovery Act of 1976 (RCRA). Employers conducting operations at treatment, storage, and disposal (TSD) facilities must develop and implement a written safety and health program for employees involved in hazardous waste operations that shall be available for inspection by employees, their representatives, and OSHA personnel. The program is to be designed to identify, evaluate, and control safety and health hazards at their facilities for the purpose of employee protection, to provide for emergency response, and to address, as appropriate, site analysis, engineering controls, maximum exposure limits, hazardous waste handling procedures, and uses of new technologies. The employer must implement a hazard communication program as part of the employer's safety and health program.

The employer must develop and implement a medical surveillance program, a decontamination procedure, a procedure for introducing new and innovative equipment into the workplace, a program for handling drums or containers, and a training program which is part of the employer's safety and health program for employees exposed to health hazards or hazardous substances at TSD operations to enable the employees to perform their assigned duties and functions in a safe and healthful manner so as not to endanger themselves or other employees. The initial training is for 24 hours and refresher training shall be for 8 hours annually. Employees who have received the initial training required by this paragraph are to be given a written certificate attesting that they have successfully completed the necessary training. Employers who can show by an employee's previous work experience and/or training, can be considered to meet the previous requirements of this paragraph.

Emergency Response

Those emergency response organizations, who have developed and implemented emergency response programs for handling releases of hazardous substances, must follow the previous guidelines.

To avoid duplications, emergency response organizations may use the local emergency response plan, the state emergency response plan, or both, as part of their emergency response plan. Those items of the emergency response plan that are being properly addressed by the SARA Title III plans may be substituted into their emergency plan, or otherwise kept together for the employer and employee's use. The senior emergency response official, responding to an emergency, becomes the individual in charge of a site-specific Incident Command System (ICS). All emergency responders and their communications are to be coordinated and controlled through the individual in charge of the ICS and assisted by the senior official present for each employer.

The individual in charge of the ICS must identify, to the extent possible, all hazardous substances or conditions present, and address, as appropriate, site analysis, use of engineering controls, maximum exposure limits, hazardous substance handling procedures, and any new technologies. Based on the hazardous substances and/or conditions present, the individual in charge of the ICS is to implement the appropriate emergency operations, and assure that the personal protective equipment worn is appropriate for the hazards to be encountered.

Employees engaged in emergency response who are exposed to hazardous substances which present an inhalation hazard, or a potential inhalation hazard, must wear positive pressure self-contained breathing apparatus until such time as the individual in charge of the ICS determines, through the use of air monitoring, that a decreased level of respiratory protection will not result in hazardous exposures to employees.

The individual in charge of the ICS must limit the number of emergency response personnel at the emergency site, in those areas of potential or actual exposure to incident or site hazards, to those who are actively performing emergency operations. However, operations in hazardous areas shall be performed using the buddy system in groups of two or more.

Back-up personnel must stand by with equipment, and be ready to provide assistance or rescue. Advance first aid support personnel, as a minimum, also stand by with medical equipment and transportation capability.

The individual in charge of the ICS may designate a safety official, who is knowledgeable in the operations being implemented at the emergency response site, with the specific responsibility of identifying and evaluating hazards and providing direction with respect to the safety of the operations for the emergency at hand. When activities are judged by the safety official to be an IDLH condition and/or to involve an imminent danger condition, the safety official has the authority to alter, suspend, or terminate those activities. The safety official is to immediately inform the individual in charge of the ICS of any actions needed to be taken to correct these hazards at the emergency scene. After emergency operations have terminated, the individual in charge of the ICS must implement appropriate decontamination procedures.

Personnel, not necessarily an employer's own employees, who are skilled in the operation of certain equipment, such as mechanized earth moving or digging equipment, or crane and hoisting equipment, and who are needed temporarily to perform immediate emergency support work that cannot reasonably be performed in a timely fashion by an employer's own employees, and who will be or may be exposed to the hazards at an emergency response scene, are not required to meet the training required for the employer's regular employees. However, these personnel are to be given an initial briefing at the site prior to their participation in any emergency response. The initial briefing must include instruction in the wearing of appropriate personal protective equipment, what chemical hazards are involved, and what duties are to be performed. All other appropriate safety and health precautions provided to the employer's own employees are to be used to assure the safety and health of these personnel.

Employees who in the course of their regular job duties work with and are trained in the hazards of specific hazardous substances, and who will be called upon to provide technical advice or assistance, at a hazardous substance release incident, to the individual in charge, must, annually, receive training, or demonstrate competency in the area of their specialization.

Training is to be based on the duties and function to be performed by each responder of an emergency response organization. The skill and knowledge levels required for all new responders, those hired after the effective date of this standard, shall be conveyed to them through training before they are permitted to take part in actual emergency operations on an incident.

First responders, at the awareness level, are individuals who are likely to witness or discover a hazardous substance release, and who have been trained to initiate an emergency response sequence by notifying the proper authorities of the release. They would take no further action beyond notifying the authorities of the release. First responders, at the awareness level, shall have sufficient training, or have had sufficient experience to objectively demonstrate competency in the following areas:

1.   An understanding of what hazardous substances are, and the risks associated with them in an incident.

2.   An understanding of the potential outcomes associated with an emergency created when hazardous substances are present.

3.   The ability to recognize the presence of hazardous substances in an emergency.

4. The ability to identify the hazardous substances, if possible.

5. An understanding of the role of the first responder awareness individual in the employer's emergency response plan, including site security and control, and the U.S. Department of Transportation's Emergency Response Guidebook.

6. The ability to realize the need for additional resources, and to make appropriate notifications to the communication center.

First responders, at the operations level, are individuals who respond to releases or potential releases of hazardous substances, as part of the initial response to the site, for the purpose of protecting nearby persons, property, or the environment from the effects of the release. They are trained to respond in a defensive fashion without actually trying to stop the release. Their function is to contain the release from a safe distance, keep it from spreading, and prevent exposures. First responders, at the operational level, must receive at least eight hours of training, or have had sufficient experience to objectively demonstrate competency in the following areas, in addition to those listed for the awareness level, and the employer shall so certify:

1. Knowledge of the basic hazard and risk assessment techniques.

2. Know of how to select and use proper personal protective equipment provided to the first responder, operational level.

3. An understanding of basic hazardous materials terms.

4. Know how to perform basic control, containment, and/or confinement operations within the capabilities of the resources and personal protective equipment available with their unit.

5. Know how to implement basic decontamination procedures.

6. An understanding of the relevant standard operating and termination procedures.

Hazardous materials technicians are individuals who respond to releases, or potential releases, for the purpose of stopping the release. They assume a more aggressive role than a first responder at the operations level in that they will approach the point of release in order to plug, patch, or otherwise stop the release of a hazardous substance. Hazardous materials technicians must have received at least 24 hours of training, equal to the first responder operations level, and in addition, must have competency in the following areas, and the employer shall so certify:

1. Know how to implement the employer's emergency response plan.

2. Know the classification, identification, and verification of known and unknown materials by using field survey instruments and equipment.

3. Be able to function within an assigned role in the Incident Command System.

4. Know how to select and use proper specialized chemical personal protective equipment provided to the hazardous materials technician.

5. Understand hazard and risk assessment techniques.

6. Be able to perform advance control, containment, and/or confinement operations within the capabilities of the resources and personal protective equipment available with the unit.

7. Understand and implement decontamination procedures.

8. Understand termination procedures.

9. Understand basic chemical and toxicological terminology and behavior.

Hazardous materials specialists are to be individuals who respond with, and provide support to hazardous materials technicians. Their duties parallel those of the hazardous materials technician, however, those duties require a more directed or specific knowledge of the various substances they may be called upon to contain. The hazardous materials specialist also acts as the site liaison with federal, state, local, and other government authorities in regard to site activities. Hazardous materials specialists must have received at least 24 hours of training equal to the technician level, and, in addition, must have competency in the following areas, and the employer shall so certify:

1. Know how to implement the local emergency response plan.

2. Understand classification, identification, and verification of known and unknown materials by using advanced survey instruments and equipment.

3. Know of the state emergency response plan.

4. Be able to select and use proper specialized chemical personal protective equipment provided to the hazardous materials specialist.

5. Understand in-depth hazard and risk techniques.

6. Be able to perform specialized control, containment, and/or confinement operations within the capabilities of the resources and personal protective equipment available.

7. Be able to determine and implement decontamination procedures.

8. Have the ability to develop a site safety and control plan.

9. Understand chemical, radiological, and toxicological terminology and behavior.

Incident commanders, who will assume control of the incident scene beyond the first responder awareness level, must receive at least 24 hours of training equal to the first responder operations level, and, in addition, must have competency in the following areas, and the employer shall so certify:

1. Know and be able to implement the employer's incident command system.

2. Know how to implement the employer's emergency response plan.

3. Know and understand the hazards and risks associated with employees working in chemical protective clothing.

4. Know how to implement the local emergency response plan.

5. Know of the state emergency response plan and of the Federal Regional Response Team.

6. Know and understand the importance of decontamination procedures.

Emergency Response Training

Trainers who teach any of the previous training subjects must have satisfactorily completed a training course for teaching the subjects they are expected to teach, such as the courses offered by the U.S. National Fire Academy; or they must have the training and/or academic credentials and instructional experience necessary to demonstrate competent instructional skills. They must also have a good command of the subject matter of the courses they are to teach.

Trainers and trainees must receive annual refresher training of sufficient content and duration to maintain their competencies, or demonstrate, annually, their competency in those areas. A statement is to be made of the training or competency, and if a statement of competency is made, the employer must keep a record of the methodology used to demonstrate competency.

Members of an organized and designated HAZMAT team, and hazardous materials specialists, must receive a baseline physical examination and be provided with medical surveillance. Any emergency response employee who exhibits signs or symptoms which may have resulted from exposure to hazardous substances during the course of an emergency incident is to be provided with medical consultation.

Upon completion of the emergency response, if it is determined that it is necessary to remove hazardous substances, health hazards, and contaminated materials (such as contaminated soil or other elements of the natural environment) from the site of the incident, the employer conducting the clean-up must comply with the standard guidelines and rules.

It is very important to refer to the referenced information found in the tables, figures, and appendices which appear in 29 CFR 1916.65, since they are too lengthy to be included in this text.

Some of the information in the appendices include personal protective equipment test methods, general descriptions and discussion of the levels of protection and protective gear, compliance guidelines, references, and training curriculum guidelines.

HEAD PROTECTION (1926.100)

The use of protective hardhats is a must on a construction jobsite. They are not just for protection from falling objects, but for all potential head injuries. The dynamic nature of a construction worksite makes bumping into objects, or being hit by flying objects, a real risk for lateral impact injuries. Other types of potential head injuries can also occur from such things as hot materials or electrical hazards.

Hardhats must meet, according to the OSHA standard, the requirements ANSI standard Z89.1-1969 for impact, and Z89.2-1997 for burns and electricity. Class A hardhats are to protect workers from impact of falling objects and low voltage electricity. Class B hardhats are to protect workers from impact and high voltage electricity. Class C hardhats are to protect workers from impact of falling objects.

There is a new ANSI standard for hardhats, Z89.1-1997, which updates the previous standard and changes the A,B,C classification to G for general, E for electrical, and C for conductive. Also, guidelines on chin straps are incorporated. Chin straps are becoming more acceptable and are used in the United States.

Hardhats are ideal for attaching other types of PPE such as ear muffs, chin straps, face shields, cold weather protection, and welding hoods (see Figure 58).

HEARING PROTECTION – OCCUPATIONAL NOISE EXPOSURE (1926.52)

Protection against the effects of noise exposure is to be provided when the sound levels exceed those shown in Table 11 of this section and are measured on the A-scale of a standard sound level meter at slow response. When employees are subjected to sound levels exceeding those listed in Table 11 of this section, feasible administrative or engineering controls are to be utilized. If such controls fail to reduce sound levels within the levels of the table, personal protective equipment is to be provided and used to reduce sound levels to within the levels of the table. If the variations in noise level involves the maxims at intervals of one second or less, it is to be considered continuous. In all cases where the sound levels exceed the

Figure 58. Hardhat with hearing protection

values shown as continuing, an effective hearing conservation program is to be administered.

When the daily noise exposure is composed of two or more periods of noise exposure of different levels, their combined effect should be considered, rather than the individual effect of each. Exposure to different levels of noise for various periods of time is to be computed using the formula in 1926.52. Exposure to impulsive or impact noise should not exceed 140 dB of peak sound pressure level.

HEATING DEVICES (TEMPORARY) (1926.154)

Temporary heating devices often cause carbon monoxide or high levels of carbon

Table 11

Permissible Noise Exposure
Courtesy of OSHA

Duration per day (Hours)	Sound level dBA slow response
8	90
6	92
4	95
3	97
2	100
1 1/2	102
1	105
1/2	110
1/4 or less	115

dioxide. Thus, the work environment must be supplied with sufficient quantities of fresh air, and the heater itself should be kept ten feet from combustible materials (see Figure 59). These heaters should be protected from tipping over, and at no time should the safety cutoff be over ridden. Also, care should be taken to insulate the heater from combustible floors. Solid fuel salamanders are prohibited in buildings and on scaffolds.

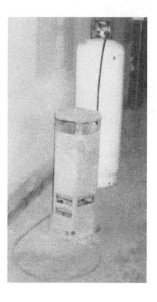

Figure 59. Example of a temporary heater

HEAVY EQUIPMENT, PREVENTING SLIPS AND FALLS

In preventing slips and falls from heavy equipment, the following precautions should be undertaken:

1. Always dismount equipment while facing ladders or steps.

2. Always make sure that three points of contact are used at all times (Example: two feet and one hand, or two hands and one foot) (see Figure 60).

3. Never jump from a piece of equipment.

4. Make sure to have proper foot wear; make sure the foot wear is clear of mud and debris.

5. After operating a piece of equipment for a period of time in the sitting position, muscles may stiffen up and may not be as flexible when dismounting equipment.

6. Use the provided hand holds, ladders, and steps for mounting and dismounting equipment.

7. Be careful when ice or snow exists on surfaces.

8. Cleaning, oiling, fueling, or repairing is not to be done while equipment is operating.

Figure 60. Construction worker mounting heavy equipment using three points of contact

HELICOPTERS (1926.551)

The use of helicopters for lifting requires special procedures, an adequately trained ground crew, and an awareness of the unique hazards involved with their use. Helicopter cranes are expected to comply with any applicable regulations of the Federal Aviation Administration. Prior to each day's operation, a briefing is to be conducted. This briefing must set forth the plan of operation for the pilot and ground personnel.

All loads must be properly slung. Taglines should be of a length that will not permit their being drawn up into rotors. Pressed sleeves, swedged eyes, or equivalent means are to be used for all freely suspended loads in order to prevent hand splices from spinning open, or cable clamps from loosening.

All electrically operated cargo hooks must have electrical activating devices so designed and installed as to prevent inadvertent operation. In addition, cargo hooks are to be equipped with an emergency mechanical control for releasing the load. The hooks are to be tested prior to each day's operation to determine that the release functions properly, both electrically and mechanically.

Workers receiving the load must have personal protective equipment which consists of complete eye protection, and hardhats secured by chinstraps. Loose-fitting clothing which is likely to flap in the downwash and be snagged on a hoist line, is not to be worn. Every practical precaution is to be taken to protect the employees from flying objects in the rotor downwash. All loose gear within 100 feet of the place of lifting the load, depositing the load, and all other areas susceptible to rotor downwash, are to be secured or removed. Good housekeeping is to be maintained in all helicopter loading and unloading areas. The helicopter operator/pilot is to be responsible for the size, weight, and manner in which loads are connected to the helicopter. Open fires are not permitted in an area that could result in such fires being spread by the rotor downwash. If, for any reason, the helicopter operator believes the lift cannot be made safely, the lift is not made. The weight of an external load must not exceed the manufacturer's rating.

When employees are required to perform work under hovering craft, a safe means of access is to be provided for employees to reach the hoist line hook, and engage or disengage cargo slings. Employees do not perform work under hovering craft except when necessary to hook or unhook loads.

Static charge on the suspended load is to be dissipated with a grounding device before ground personnel touch the suspended load, or protective rubber gloves are to be worn by all ground personnel touching the suspended load.

Ground lines, hoist wires, or other gear, except for pulling lines or conductors that are allowed to "pay out" from a container or roll off a reel, are not to be attached to any fixed ground structure, or allowed to foul on any fixed structure.

When visibility is reduced by dust or other conditions, ground personnel must exercise special caution to keep clear of main and stabilizing rotors. Precautions are also to be taken by the employer to eliminate, as far as practical, reduced visibility. Signal systems between aircrew and ground personnel must be understood and checked in advance of hoisting the load. This applies to either radio or hand signal systems. Hand signals are as shown in Figure 61.

No unauthorized person is allowed to approach within 50 feet of the helicopter when the rotor blades are turning. Whenever approaching or leaving a helicopter with blades rotating, all employees are to remain in full view of the pilot and keep in a crouched position. Employees should avoid the area from the cockpit or cabin rearward unless authorized by the helicopter operator to work there.

For safe helicopter loading and unloading operations, sufficient ground personnel needs to be provided. There must be constant, reliable communication between the pilot and the designated employee, of the ground crew, who is acting as the signalman during the period of loading and unloading. This signalman is to be distinctly recognizable from all other ground personnel.

HOIST, BASE-MOUNTED DRUM (1926.553)

Base-mounted drum hoists, exposed to moving parts, such as gears, projecting screws, setscrews, chains, cables, chain sprockets, and reciprocating or rotating parts, which constitute a hazard, must be guarded. All controls used during the normal operation cycle are to be located within easy reach of the operator's station. Electric motor-operated hoists are to be provided with a device to disconnect all motors from the line, upon power failure, and no motor is permitted to be restarted until the controller handle is brought to the "off" position. And where applicable, an overspeed preventive device is to exist. Also, a means must be present whereby a remotely operated hoist stop can be activated when any control is ineffective. All base-mounted drum hoists in use must meet the applicable requirements for design, construction, installation, testing, inspection, maintenance, and operations, as prescribed by the manufacturer.

HOIST, MATERIALS   (1926.552)

All material hoists must comply with the manufacturer's specifications and limitations, which are applicable to the operation of all hoists and elevators. Where manufacturer's specifications are not available, the limitations assigned to the equipment are to be based on the determinations of a professional engineer competent in the field. Rated load capacities, recommended operating speeds, and special hazard warnings or instructions are to be posted on cars and platforms.

Hoisting ropes are to be installed in accordance with the wire rope manufacturers' recommendations. Wire rope is to be removed from service when any of the following conditions exist:

Figure 61. Hand signals used by helicopter ground personnel. Courtesy of OSHA from Figure N-1 of 29 CFR 1926.551.

1.   In hoisting ropes, six randomly distributed broken wires in one rope lay, or three broken wires in one strand in one rope lay.

2.   Abrasion, scrubbing, flattening, or peening, causing loss of more than one-third of the original diameter of the outside wires.

3.   Evidence of any heat damage resulting from a torch, or any damage caused by contact with electrical wires.

4.   Reduction from nominal diameter of more than three sixty-fourths inch for diameters up to and including three-fourths inch; one-sixteenth inch for diam-

eters seven-eights to 1 1/8 inches; and three thirty-seconds inch for diameters 1 1/4 to 1 1/2 inches.

The installation of live booms on hoists is prohibited. The use of endless belt-type manlifts on construction is prohibited. Operating rules for material hoists are to be established and posted at the operator's station of the hoist. Such rules include signal systems and the allowable line speed for various loads. Rules and notices are to be posted on the car frame or crosshead in a conspicuous location, including the statement "No Riders Allowed." No person is allowed to ride on material hoists except for the purposes of inspection and maintenance.

All entrances of the hoistways are to be protected by substantial gates or bars which guard the full width of the landing entrance. All hoistway entrance bars and gates are to be painted with diagonal contrasting colors, such as black and yellow stripes. Bars must not be less than 2- by 4-inch wooden bars, or the equivalent, located two feet from the hoistway line. Bars are to be located not less than 36 inches, nor more than 42 inches above the floor. Gates or bars protecting the entrances to hoistways are to be equipped with a latching device.

An overhead protective covering of 2-inch planking, 3/4-inch plywood, or other solid material of equivalent strength, is to be provided on the top of every material hoist cage or platform. The operator's station of a hoisting machine is to be provided with overhead protection equivalent to tight planking not less than 2 inches thick. The support for the overhead protection is to be of equal strength.

Hoist towers may be used with or without an enclosure on all sides. However, whichever alternative is chosen, the following applicable conditions are to be met:

1. When a hoist tower is enclosed, it is to be enclosed on all sides, for its entire height, with a screen enclosure of 1/2-inch mesh, No. 18 U.S. gauge wire, or equivalent, except for landing access.

2. When a hoist tower is not enclosed, the hoist platform or car is to be totally enclosed (caged) on all sides, for the full height between the floor and the overhead protective covering, with 1/2-inch mesh of No. 14 U.S. gauge wire or equivalent. The hoist platform enclosure must include the required gates for loading and unloading. A 6-foot high enclosure is to be provided on the unused sides of the hoist tower at ground level.

Car arresting devices are to be installed to function in case of rope failure. All material hoist towers are to be designed by a licensed professional engineer. All material hoists must conform to the requirements of ANSI A10.5-1969, Safety Requirements for Material Hoists.

HOIST, OVERHEAD (1926.554)

When operating an overhead hoist, the safe working load, as determined by the manufacturer, is to be indicated on the hoist, and this safe working load shall not be exceeded. Also, the supporting structure to which the hoist is attached must have a safe working load equal to that of the hoist. The support is to be arranged so as to provide for free movement of the hoist, and must not restrict the hoist from lining itself up with the load. Overhead hoists are to be installed only in locations that will permit the operator to stand clear of the load at all times.

An air driven hoist must be connected to an air supply of sufficient capacity and pressure in order to safely operate the hoist. All air hoses supplying air are to be positively connected to prevent their becoming disconnected during use.

All overhead hoists in use must meet the applicable requirements for construction, design, installation, testing, inspection, maintenance, and operation, as prescribed by the manufacturer.

HOIST, PERSONNEL

Personnel hoists/hoist towers outside the structure are to be enclosed for the full height on the side, or the sides used for entrance and exit to the structure. At the lowest landing, the enclosures on the sides which are not being used for exit or entrance to the structure, must be enclosed to a height of at least 10 feet. Other sides of the tower which are adjacent to floors or scaffold platforms are to be enclosed to a height of 10 feet above the level of such floors or scaffolds. Towers inside of structures are to be enclosed on all four sides throughout the full height.

Towers are to be anchored to the structure at intervals not exceeding 25 feet. In addition to tie-ins, a series of guys are installed. Where tie-ins are not practical, the tower is to be anchored by means of guys made of wire rope at least one-half inch in diameter, securely fastened to anchorage to ensure stability.

Hoistway doors or gates are not to be less than 6 feet 6 inches high, and shall be provided with mechanical locks which cannot be operated from the landing side; they are be accessible only to persons on the car.

Cars are to be permanently enclosed on all sides and the top, except sides used for entrance and exit which have car gates or doors. A door or gate is to be provided at each entrance to the car which protects the full width and height of the car entrance opening. An overhead protective covering of 2-inch planking, 3/4-inch plywood, or other solid material or equivalent strength, is to be provided on the top of every personnel hoist. Doors or gates must be provided with electric contacts which do not allow movement of the hoist when the door or gate is open. Safeties are to be capable of stopping and holding the car and rated load when traveling at governor tripping speed. Cars are to be provided with a capacity and data plate, secured in a conspicuous place on the car or crosshead. Internal combustion engines are not permitted for direct drive. Normal and final terminal stopping devices are to be provided, and an emergency stop switch is to be provided in the car and marked "Stop."

The minimum number of hoisting ropes used is: three for traction hoists and two for drum-type hoists. The minimum diameter of hoisting and counterweight wire ropes is 1/2-inch. Minimum factors of safety for suspension wire ropes can be found in 29 CFR 1926.552.

Following assembly and erection of hoists, and before being put in service, inspection and testing of all functions and safety devices are to be made, under the supervision of a competent person. A similar inspection and test is required following a major alteration of an existing installation. All are to be inspected and tested at not more than 3-month intervals. The employer must prepare a certification record which includes the date the inspection and test of all functions and safety devices was performed; the signature of the person who performed the inspection and test; and a serial number, or other identifier, for the hoist that was inspected and tested. The most recent certification record is to be maintained on file.

All personnel hoists used by employees are to be constructed of materials and components which meet the specifications for materials, construction, safety devices, assembly, and structural integrity as stated in the American National Standard A10.4-1963, Safety Requirements for Workmen's Hoists. Personnel hoists used in bridge tower construction must be approved by a registered professional engineer and erected under the supervision of a qualified engineer competent in this field.

When a hoist tower is not enclosed, the hoist platform or car is to be totally enclosed (caged) on all sides, for the full height between the floor and the overhead protective covering, with 3/4-inch mesh of No. 14 U.S. gauge wire or equivalent. The hoist platform enclosure must include the required gates for loading and unloading.

These hoists are to be inspected and maintained on a weekly basis. Whenever the hoisting equipment is exposed to winds exceeding 35 miles per hour, it is to be inspected and put in operable condition before reuse.

Wire rope is to be taken out of service when any of the following conditions exist:

1. In running ropes, six randomly distributed broken wires in one lay, or three broken wires in one strand in one lay.

2. Wear of one-third the original diameter of outside individual wires, kinking, crushing, bird caging, or any other damage resulting in distortion of the rope structure.

3. Evidence of any heat damage from any cause.

4. Reductions from nominal diameter of more than three-sixty-fourths inch for diameters to, and including three-fourths inch, one-sixteenth inch for diameters seven-eights inch to 1 1/8 inches inclusive, and three-thirty-seconds inch for diameters 1 1/4 to 1 1/2 inches inclusive.

5. In standing ropes, more than two broken wires in one lay in sections beyond end connections, or more than one broken wire at an end connection.

Permanent elevators, under the care and custody of the employer, and used by employees for work covered by this Act, shall comply with the requirements of American National Standards Institute A17.1-1965 with addenda A17.1a-1967, A17.1b-1968, A17.1c-1969, A17.1d-1970, and must be inspected in accordance with A17.2-1960 with addenda A17.2a-1965, A17.2b-1967.

HOUSEKEEPING (1926.25)

Housekeeping means keeping everything at work in its proper place and putting things away after they are used. Tools and materials should never be left on the floor, on stairs, in walkways, or aisles. During the course of construction, alteration, or repairs, form and scrap lumber with protruding nails, and all other debris, are to be kept cleared from work areas, passageways, and stairs, in and around buildings or other structures. Combustible scraps and debris are to be removed at regular intervals during the course of construction; a safe means of facilitating such removal must be provided. (See Figure 62.)

Containers are to be provided for the collection and separation of waste, trash, oily and used rags, and other refuse. Containers used for garbage and other oily, flammable, or hazardous wastes, such as caustics, acids, harmful dusts, etc. are to be equipped with covers. Garbage and other waste must be disposed of at frequent and regular intervals.

ILLUMINATION (1926.56)

Determining the adequacy of illumination for an ongoing construction worksite is not always an easy process. Employers should respond to workers' complaints of inadequate lighting; this may be the best indicator of adequate lighting. Although construction areas, ramps, runways, corridors, offices, shops, and storage areas are to meet the minimums found in Table D-3 of 29 CFR 1926.56; actual measurements are seldom done on construction sites. Thus, prudent judgment should be used when evaluating the amount of illumination necessary. This

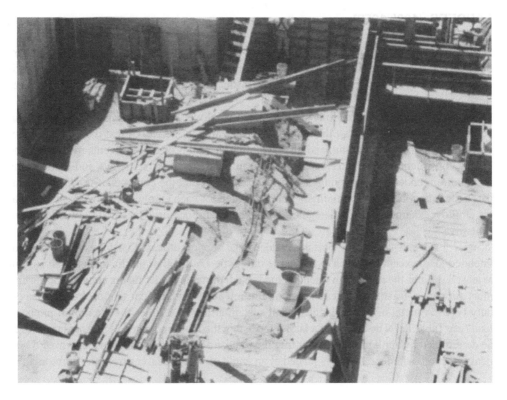

Figure 62. Typical housekeeping problems that arise on construction sites

is especially true since it is widely recognized that poorly illuminated areas result in slips and falls which lead to other injuries.

JACKS (1926.305)

When using jacks, the manufacturer's rated capacity must be legibly marked on all jacks and not exceeded. All jacks should have a positive stop to prevent overtravel. When it is necessary to provide a firm foundation, the base of the jack is to be blocked or cribbed. Where there is a possibility of slippage of the metal cap of the jack, a wood block is to be placed between the cap and the load. After the load has been raised, the load must be cribbed, blocked, or otherwise secured at once.

Hydraulic jacks which are exposed to freezing temperatures must be supplied with an adequate antifreeze liquid. All jacks need to be properly lubricated at regular intervals. Each jack must be thoroughly inspected, at times, depending upon the service conditions. Inspections are not to be less frequent than the following:

1. For constant or intermittent use at one locality—once every 6 months.

2. For jacks sent out of shop for special work—when sent out, and when returned.

3. For a jack subjected to abnormal load or shock—immediately before, and immediately thereafter.

Repair or replacement parts are to be examined for possible defects. Jacks which are out of order are to be tagged accordingly, and are not to be used until repairs are made.

LADDERS (1926.1053)

The employer must, as necessary, provide training for each employee using ladders and stairways. The program should enable each employee to recognize hazards related to ladders and stairways, and train each employee in the procedures to be followed to minimize these hazards.

The employer needs to ensure that each employee has been trained by a competent person in the following areas: the nature of fall hazards in the work area; the correct procedures for erecting, maintaining, and disassembling the fall protection systems to be used; the proper construction, use, placement, and care in handling of all stairways and ladders; the maximum intended load-carrying capacities of ladders; and the contents of the OSHA standards regarding ladders. Retraining is to be provided for each employee, as necessary, so that the employee maintains the understanding and knowledge acquired for safe use of ladders and stairways.

Each self-supporting portable ladder must support at least four times the maximum intended load, except that each extra-heavy-duty type 1A metal or plastic ladder shall sustain at least 3.3 times the maximum intended load. Each portable ladder that is not self-supporting must support at least four times the maximum intended load, except that each extra-heavy-duty type 1A metal or plastic ladders shall sustain at least 3.3 times the maximum intended load (see Table 12). The ability of a ladder to sustain the loads indicated in this paragraph is to be determined by applying or transmitting the requisite load to the ladder in a downward vertical direction, when the ladder is placed at an angle of 75 1/2 degrees from the horizontal.

Table 12

Ladder Types and Load Capacities

Type	Grade	Duty Rating
III	Household	200 lbs.
II	Commercial	225 lbs.
I	Industrial	250 lbs.
IA	Extra Heavy Duty Industrial	300 lbs

Ladder rungs, cleats, and steps shall be parallel, level, and uniformly spaced when the ladder is in position for use. Rungs, cleats, and steps of portable and fixed ladders are to be spaced not less than 10 inches apart, nor more than 14 inches apart, as measured between center lines of the rungs, cleats and steps. Rungs, cleats, and steps of step stools are not to be less than 8 inches apart, nor more than 12 inches apart, as measured between center lines of the rungs, cleats, and steps. Rungs, cleats, and steps of the base section of extension trestle ladders are not to be less than 8 inches, nor more than 18 inches apart, as measured between center lines of the rungs, cleats, and steps. The rung spacing on the extension section of the extension trestle ladder must not be less than 6 inches, nor more than 12 inches, as measured between center lines of the rungs, cleats, and steps. The rungs of individual-rung/step ladders are to be shaped such that the employees' feet cannot slide off the end of the rungs. The rungs and steps of portable metal ladders must be corrugated, knurled, dimpled, coated with skid-resistant material, or otherwise treated to minimize slipping.

The minimum clear distance between the sides of individual-rung/step ladders, and the minimum clear distance between the side rails of other fixed ladders, is to be 16 inches. The minimum clear distance between side rails for all portable ladders is to be 11 1/2 inches.

Ladders are not to be tied or fastened together to provide longer sections, unless they are specifically designed for such use. A metal spreader or locking device is to be provided, on each stepladder, to hold the front and back sections in an open position when the ladder is being used (see Figure63). When splicing is required to obtain a given length of side rail, the resulting side rail must be at least equivalent in strength to a one-piece side rail made of the same material.

Figure 63. The safe use of a stepladder is standing no higher than the second step from the top

Except when portable ladders are used to gain access to fixed ladders (such as those on utility towers, billboards, and other structures where the bottom of the fixed ladder is elevated to limit access), or when two or more separate ladders are used to reach an elevated work area, the ladders are to be offset with a platform or landing between the ladders.

Ladder components are to be surfaced so as to prevent injury to an employee from punctures or lacerations, and to prevent snagging of clothing. Wood ladders are not to be coated with any opaque covering, except for identification or warning labels which may be placed on one face, only, of a side rail.

The minimum perpendicular clearance between fixed ladder rungs, cleats, and steps, and any obstruction behind the ladder shall be 7 inches, except in the case of an elevator pit ladder for which a minimum perpendicular clearance of 4 1/2 inches is required. The minimum perpendicular clearance between the center line of fixed ladder rungs, cleats, and steps, and any obstruction on the climbing side of the ladder shall be 30 inches, except as provided in paragraph (a)(15) of this section. When unavoidable obstructions are encountered, the minimum perpendicular clearance between the centerline of fixed ladder rungs, cleats, and steps, and the obstruction on the climbing side of the ladder may be reduced to 24 inches, provided that a deflection device is installed to guide employees around the obstruction.

Fixed Ladders

Each fixed ladder must support at least two loads of 250 pounds each, concentrated between any two consecutive attachments (the number and position of additional concentrated loads of 250 pounds each, determined from anticipated usage of the ladder, are also included), plus anticipated loads caused by ice buildup, winds, and rigging, and impact loads resulting from the use of ladder safety devices. Each step or rung shall be capable of supporting a single concentrated load of a least 250 pounds applied in the middle of the step or rung.

Through-fixed ladders at their point of access/egress must have a step-across distance of not less than 7 inches, nor more than 12 inches, as measured from the centerline of the steps or rungs, to the nearest edge of the landing area. If the normal step-across distance exceeds 12 inches, a landing platform shall be provided to reduce the distance to the specified limit. Fixed ladders are to be used at a pitch no greater than 90 degrees from the horizontal, as measured to the back side of the ladder. Fixed ladders, without cages or wells, shall have a clear width of at least 15 inches on each side of the centerline of the ladder, to the nearest permanent object. Fixed ladders are to be provided with cages, wells, ladder safety devices, or self-retracting lifelines where the length of climb is less than 24 feet, but the top of the ladder is to be at a distance greater than 24 feet above lower levels. Where the total length of a climb equals or exceeds 24 feet, fixed ladders shall be equipped with one of the following:

1. Ladder safety devices.

2. Self-retracting lifelines and rest platforms at intervals not to exceed 150 feet.

3. A cage or well and multiple ladder sections and each ladder section must not exceed 50 feet in length. Ladder sections shall be offset from adjacent sections, and landing platforms shall be provided at maximum intervals of 50 feet.

Cages for fixed ladders shall conform to all of the following horizontal bands and shall be fastened to the side rails of rail ladders, or fastened directly to the structure, building, or equipment. For individual-rung ladders, vertical bars are to be on the inside of the horizontal bands and shall be fastened to them. The cages shall not extend less than 27 inches, or more than 30 inches from the centerline of the step or rung (excluding the flare at the bottom of the cage), and must not be less than 27 inches in width. The inside of the cage is to be clear of projections and horizontal bands are to be spaced not more than 4 feet on center vertically. The vertical bars are to be spaced at intervals not more than 9 1/2 inches on center horizontally and the bottom of the cage is to be at a level not less than 7 feet, nor more than 8 feet above the point of access to the bottom of the ladder. The bottom of the cage shall be flared not less than 4 inches all around within the distance between the bottom horizontal band and the next higher band. The top of the cage is to be a minimum of 42 inches above the top of the platform, or the point of access at the top of the ladder, with provision for access to the platform or other point of access.

Wells for fixed ladders must conform to all of the following: they must completely encircle the ladder; they are to be free of projections; the inside face on the climbing side of the ladder must not extend less than 27 inches, nor more than 30 inches from the centerline of the step or rung. The inside clear width is to be at least 30 inches, and the bottom of the wall, on the access side, must start at a level not less than 7 feet, nor more than 8 feet above the point of access to the bottom of the ladder.

Ladder safety devices, and related support systems, for fixed ladders must conform to all of the following: they shall be capable of withstanding, without failure, a drop test consisting of an 18-inch drop of a 500-pound weight. They must permit the employee using the device to ascend or descend without continually having to hold, push, or pull any part of the device, leaving both hands free for climbing. They are to be activated within 2 feet after a fall occurs, and limit the descending velocity of an employee to 7 feet/sec. or less. The connection between the carrier or lifeline, and the point of attachment to the body harness, must not exceed 9 inches in length.

The mounting of ladder safety devices for fixed ladders must conform to the following: mountings for rigid carriers are to be attached at each end of the carrier with intermediate mountings, as necessary, and spaced along the entire length of the carrier in order to provide the strength necessary to stop employees' falls. Mountings for flexible carriers are to be attached at each end of the carrier. When the system is exposed to wind, cable guides for flexible carriers shall be installed at a minimum spacing of 25 feet, and a maximum spacing of 40 feet along the entire length of the carrier in order to prevent wind damage to the system. The design and installation of mountings and cable guides must not reduce the design strength of the ladder.

The side rails of through or side-step fixed ladders must extend 42 inches above the top of the access level or landing platform served by the ladder. For a parapet ladder the access is the top of the roof if the parapet is cut to allow access then the access level shall be the top of the parapet. For through-fixed ladder extensions, the steps or rungs are to be omitted from the extension, and the extension of the side rails is to be flared to provide not less than 24 inches, nor more than 30 inches clearance between side rails. Where ladder safety devices are provided, the maximum clearance between side rails of the extensions must not exceed 36 inches. For side-step fixed ladders, the side rails and the steps or rungs are to be continuous in the extension. Individual-rung/step ladders, except those used where their access openings are covered with manhole covers or hatches, must extend at least 42 inches above an access level or landing platform by either the continuation of the rung spacing as horizontal grab bars, or by providing vertical grab bars that have the same lateral spacing as the vertical legs of the rungs.

Rules for All Ladders

Rules for the use of all ladders, including job-made ladders, are when portable ladders are used for access to an upper landing surface, the ladder's side rails shall extend at least 3 feet above the upper landing surface to which the ladder is used to gain access; or when such an extension is not possible because of the ladder's length, then the ladder shall be secured, at its top, to a rigid support that will not deflect, and a grasping device, such as a grabrail, shall be provided to assist employees in mounting and dismounting the ladder (see Figure 64). In no case shall the extension be such that ladder deflection under a load would, by itself, cause the ladder to slip off its support.

Figure 64. Extension ladder secured and three rungs above the landing area

Ladders are to be maintained free of oil, grease, and other slipping hazards. Also, ladders are not to be loaded beyond the maximum intended load for which they were built, nor beyond their manufacturer's rated capacity. Ladders must be used only for the purpose for which they were designed.

Non-self-supporting ladders are to be used at an angle such that the horizontal distance from the top support, to the foot of the ladder, is approximately one-quarter of the working length of the ladder (the distance along the ladder between the foot and the top support). Wood job-made ladders, with spliced side rails, are to be used at an angle such that the horizontal distance is one-eighth the working length of the ladder.

Ladders are to be used only on stable and level surfaces unless secured in order to prevent accidental displacement (see Figure 65); they are not to be used on slippery surfaces unless secured, or provided with slip-resistant feet which prevents accidental displacement. Slip-resistant feet shall not be used as a substitute for care in placing, lashing, or holding a ladder that is used upon slippery surfaces including, but not limited to, flat metal or concrete surfaces that are constructed so they cannot be prevented from becoming slippery.

Figure 65. Safe use of an extension ladder by using an extra person to prevent
*   displacement*

Ladders placed in any location where they can be displaced by workplace activities or traffic, such as in passageways, doorways, or driveways, are to be secured in order to prevent accidental displacement, or a barricade must be used to keep the activities or traffic away from the ladder. The area around the top and bottom of ladders is to be kept clear.

The top of a non-self-supporting ladder is to be placed with the two rails supported equally, unless it is equipped with a single support attachment. Ladders are not to be moved, shifted, or extended while occupied. Ladders must have nonconductive siderails if they are used where the employee or the ladder could contact exposed energized electrical equipment.

The top or top step of a stepladder is not to be used as a step. Cross-bracing on the rear section of a stepladder is not to be used for climbing unless the ladder is designed and provided with steps for climbing on both front and rear sections.

Ladders are to be inspected on a periodic basis by a competent person for visible defects, and after any occurrence that could affect their safe use. Portable ladders with structural defects such as, but not limited to, broken or missing rungs, cleats, steps, broken or split rails, corroded components, or other faulty or defective components, are to be either immediately marked in a manner that readily identifies them as defective, or be tagged with "Do Not Use," or similar language, and are to be withdrawn from service until repaired.

Fixed ladders with structural defects such as, but not limited to, broken or missing rungs, cleats, steps, broken or split rails, or corroded components, are to be withdrawn from service until repaired. The requirement to withdraw a defective ladder from service is satisfied if the ladder is

1. Immediately tagged with "Do Not Use" or similar language, or

2. Marked in a manner that readily identifies it as defective, or

3. Blocked (such as with a plywood attachment that spans several rungs).

Ladder repairs must restore the ladder to a condition meeting its original design criteria before the ladder is returned to use. Single-rail ladders are not to be used. When ascending or descending a ladder, the user must face the ladder and keep the user's weight centered between the rails. Each worker should maintain three points of contact when progressing up and/or down the ladder. Employees are not to carry any object or load while on a ladder that could cause the employee to lose balance and fall; also, do not move, shift, or extend the ladder while occupied. Do not use the top two steps of a step ladder, or top three rungs of extension ladder (see Figure 66). Do not use a ladder when you need a scaffold.

Figure 66. Unsafe use of a stepladder by standing on top step

It is important that job made ladders are substantially built (see Figure 67) and that the right size ladder is selected to accomplish the assigned task (see Figure 68). No single rail ladders are permitted on construction jobsites.

LIFT-SLAB CONSTRUCTION (1926.705)

Due to the 28 workers who died at L'Ambiance Plaza in April of 1987 while using lift-slab construction procedures, it was deemed that a need existed for a standard regarding this construction process. Thus, lift-slab operations are to be designed and planned by a registered professional engineer who has experience in lift-slab construction. Such plans and designs shall be implemented by the employer and include detailed instructions and sketches indicating the prescribed method of erection. These plans and designs must also include provisions for ensuring the lateral stability of the building/structure during construction.

Jacks/lifting units are to be marked to indicate their rated capacity, as established by the manufacturer. Jacks/lifting units are not to be loaded beyond their rated capacity, as established by the manufacturer. Jacking equipment is to be capable of supporting at least two and one-half times the load being lifted during jacking operations, and the equipment is not to be overloaded. For the purpose of this provision, jacking equipment includes any load bearing component which is used to carry out the lifting operation(s). Such equipment includes, but is not limited, to the following: threaded rods, lifting attachments, lifting nuts, hook-up collars, T-caps, shearheads, columns, and footings. Jacks/lifting units must be designed and installed so that they will neither lift nor continue to lift when they are loaded in excess of their rated capacity. Jacks/lifting units must have a safety device installed which will cause the jacks/ lifting units to support the load, in any position, in the event any jacklifting unit malfunctions or loses its lifting ability.

Figure 67. Job-made ladder   *Figure 68. Ladders of different lengths should be available*

Jacking operations are to be synchronized in such a manner as to ensure even and uniform lifting of the slab. During lifting, all points at which the slab is supported are to be kept within 1/2 inch of that needed to maintain the slab in a level position. If leveling is automatically controlled, a device is to be installed that will stop the operation when the 1/2 inch tolerance is exceeded, or where there is a malfunction in the jacking (lifting) system. If leveling is maintained by manual controls, such controls are to be located in a central location and attended by a competent person while the whole lift is in progress.

The maximum number of manually controlled jacks/lifting units on one slab is to be limited to a number that will permit the operator to maintain the slab level within specified tolerances, but in no case shall that number exceed 14.

No workers, except those essential to the jacking operation, are to be permitted in the building/structure while any jacking operation is taking place, unless the building/structure has been reinforced sufficiently to ensure its integrity during erection. The phrase "reinforced sufficiently to ensure its integrity" means that a registered professional engineer, independent of the engineer who designed and planned the lifting operation, has determined, from the plans, that if there is a loss of support at any jack location, that loss will be confined to that location, and the structure as a whole will remain stable. Under no circumstances is any employee, who is not essential to the jacking operation, permitted to be immediately beneath a slab while it is being lifted. A jacking operation begins when a slab, or group of slabs, is lifted, and the operation ends when such slabs are secured (with either temporary connections or permanent connections). When making temporary connections to support slabs, wedges are to be secured by tack welding, or an equivalent method of securing the wedges, in order to prevent them from falling out of position. Lifting rods may not be released until the wedges at that column have been secured. All welding on temporary and permanent connections is to be performed by a certified welder, familiar with the welding requirements, specified in the plans, and specifications for the lift-slab operation. Load transfer from jacks/lifting units to building columns is not to be executed until the welds on the column shear plates (weld blocks) are cooled to air temperature.

Jacks/lifting units are to be positively secured to building columns so that they do not become dislodged or dislocated. Equipment is to be designed and installed so that the lifting rods cannot slip out of position, or the employer must institute other measures, such as the use of locking or blocking devices, which will provide positive connection between the lifting rods and attachments, and will prevent components from disengaging during lifting operations.

LIQUID-FUEL TOOLS (1926.302)

Liquid-fuel tools are usually powered by gasoline. Vapors that can burn or explode and give off dangerous exhaust fumes are the most serious hazards associated with liquid-fuel tools. Only assigned, qualified operators shall operate power, powder-actuated, or air-driven tools. The following safe work procedures must be followed when using liquid-fuel: handle, transport, and store gas or fuel only in approved flammable liquid containers; before refilling the tank for a fuel-powered tool, the user must shut down the engine and allow it to cool to prevent accidental ignition of hazardous vapors; and effective ventilation and/or personal protective equipment must be provided when using a fuel-powered tool inside a closed area. A fire extinguisher must be readily available in the work area.

LIQUID PETROLEUM GAS (1926.153)

The fire hazard involved with LP Gas makes it imperative to follow precautions, as a preventive measure, in the handling of it. LP gas systems must have containers, valves, connectors, manifold valve assemblies, and regulators of an approved type. All cylinders must meet the Department of Transportation's specification identification requirements published in 49 CFR Part 178, Shipping Container Specifications. Welding on LP-Gas containers is prohibited. Container valves, fittings, and accessories connected directly to the container, including primary shut-off valves, must have a rated working pressure of at least 250 psig. and must be of a material and design suitable for LP-Gas service. Every container and every vaporizer is to be provided with one or more approved safety relief valves or devices. Care must be taken when filling the fuel containers of trucks or motor vehicles from bulk storage containers. This must be performed not less than 10 feet from the nearest masonry-walled building, or not less than 25 feet from the nearest building or other construction and, in any event, not less than 25 feet from any building opening. Also, the filling of portable containers or containers mounted on skids from storage containers, must not be performed less than 50 feet from the nearest building. Regulators are to be either directly connected to the container valves or to manifolds connected to the container valves. The regulator is to be suitable for use with LP-Gas. Manifolds and fittings, which connect containers to pressure regulator inlets, are to be designed for at least 250 psig. service pressure.

LP-Gas consuming appliances are to be approved types and in good condition. Aluminum piping or tubing shall not be used. Hose should be designed for a working pressure of at least 250 psig. Design, construction, and performance of hose and hose connections must have their suitability determined by listing by a nationally recognized testing agency.

Portable heaters, including salamanders, should be equipped with an approved automatic device to shut off the flow of gas to the main burner, and pilot, if used, in the event of flame failure. Container valves, connectors, regulators, manifolds, piping, and tubing are not to be used as structural supports for heaters. Containers, regulating equipment, manifolds, pipe, tubing, and hose are to be located where there will be minimized exposure to high temperatures or physical damage. If two or more heater-container units, of either the integral or nonintegral type, are located in an unpartitioned area on the same floor, the container or containers of each unit are to be separated from the container or containers of any other unit by at least 20 feet. Storage of LPG within buildings is prohibited.

Storage outside of buildings, for containers awaiting use, is to be located from the nearest building or group of buildings, in accordance with Table F-3 of 29 CFR 1926.153. Containers are to be placed in a suitable ventilated enclosure or otherwise protected against tampering. Storage locations need to be provided with at least one approved portable fire extinguisher having a rating of not less than 20-B:C. When potential damage to LP-Gas systems from vehicular traffic is a possibility, precautions against such damage are taken.

The minimum separation between a liquefied petroleum gas container, and a flammable or combustible liquid storage tank, is to be 20 feet, except in the case of flammable or combustible liquid tanks operating at pressures exceeding 2.5 psig., or equipped with emergency venting which will permit pressures to exceed 2.5 psig. Suitable means are to be taken to prevent the accumulation of flammable or combustible liquids under adjacent liquefied petroleum gas containers, such as by diversion curbs or grading. When flammable or combustible liquid storage tanks are within a diked area, the liquefied petroleum gas containers are to be outside the diked area and at least 10 feet away from the centerline of the wall of the diked area. The foregoing provisions do not apply when liquefied petroleum gas containers of 125 gallons or less capacity are installed adjacent to fuel oil supply tanks of 550 gallons, or less capacity.

LOCKOUT/TAGOUT (1910.147)

Although there is not a specific regulation for the construction industry regarding lockout/tagout, each construction company should develop a program or policy regarding lockout/tagout which covers the construction, servicing, and maintenance of machines, electrical circuits, piping containing liquids, and equipment in which the "unexpected" energization or start up of the machines or equipment, or release of stored energy, could cause injury to employees.

If a worker is required to remove or bypass a guard or other safety device, is required to place any part of his or her body into an area on a machine or piece of equipment where work is actually performed upon the material being processed (point of operation), or where an associated danger zone exists during a machine operating cycle or where the employee may contact, be engulfed, or inundated by a source of energy, then lockout/tagout procedures should be implemented.

Lockout/tagout should include work on cords and plug connected to electric equipment for which there could be exposure to the hazards of unexpected energizing or start up of the equipment. Lockout/tagout should also include hot tap operations involving transmission and distribution systems for substances such as gas, steam, water, or petroleum products when they are performed on pressurized pipelines. Another example of this is that of a mechanical energy source that needs to be locked or blocked into place, such as when a dump truck's bed is raised. Other sources of energy would include hydraulic, pneumatic, chemical, thermal, or other energy.

When such conditions exist, and in order to prevent injury to employees, employers should establish a program and utilize procedures for affixing appropriate lockout or tagout devices to the energy isolating devices, and to otherwise disabled machines or equipment, in order to prevent unexpected energization, start up, or the release of stored energy.

An energy isolating device is a mechanical device that physically prevents the transmission or release of energy, including, but not limited to, the following: a manually operated electrical circuit breaker, a disconnect switch, a manually operated switch by which the conductors of a circuit can be disconnected from all ungrounded supply conductors and, in addition, no pole can be operated independently; a line valve; a block; and any similar device used to block or isolate energy. Push buttons, selector switches, and other control circuit type devices are not to be used as energy isolating devices.

An energy isolating device must be capable of being locked out, if it has a hasp or other means of attachment to which, or through which, a lock can be affixed, or it has a locking mechanism built into it. Other energy isolating devices are to be capable of being locked out, if lockout can be achieved without the need to dismantle, rebuild, or replace the energy isolating device, or permanently alter its energy control capability.

Energy Control Program

Contractors should establish a program consisting of energy control procedures, employee training, and perform periodic inspections to ensure that before any employee does any servicing, construction, or maintenance on a machine or equipment, where the unexpected energizing, startup, or release of stored energy could occur and cause injury, the machine or equipment must be isolated from the energy source and rendered inoperative.

Although tagout systems have been used when an energy isolating device is not lockable, the only truly safe energy isolating device is the one which is lockable. Although it seems highly unlikely that the contractor can demonstrate that a tagout program will provide a level of safety equivalent to that obtained by using a lockout program, there might be isolated occasions where it might occur. Other means to be considered as part of the demonstration of full employee protection could include the implementation of additional safety measures such as the removal of an isolating circuit element, the blocking of a controlling switch, the opening of an extra disconnecting device, or the removal of a valve handle to reduce the likelihood of inadvertent energization.

Contractors should develop, use, and implement energy control procedures, for the control of potentially hazardous energy, where worker exposure could, or does exist. This may not be necessary if it can be documented that

1. The machine or equipment has no potential for stored or residual energy, or reaccumulation of stored energy after shut down which could endanger employees.

2. The machine or equipment has a single energy source which can be readily identified and isolated.

3.   The isolation and locking out of the energy source will completely de-energize and deactivate the machine or equipment.

4.   The machine or equipment is isolated from that energy source and locked out during servicing or maintenance.

5.   A single lockout device will achieve a locked-out condition.

6.   The lockout device is under the exclusive control of the authorized employee performing the servicing or maintenance.

7.   The servicing or maintenance does not create hazards for other employees.

8.   The employer, in utilizing this exception, has had no accidents involving the unexpected activation or re-energizing of the machine or equipment during servicing or maintenance.

The procedures shall clearly and specifically outline the scope, purpose, authorization, rules, and techniques to be utilized for the control of hazardous energy, and also outline the means to enforce compliance, including, but not limited to, the following:

1.   A specific statement of the intended use of the procedure;

2.   Specific procedural steps for shutting down, isolating, blocking, and securing machines or equipment to control hazardous energy;

3.   Specific procedural steps for the placement, removal, and transfer of lockout devices or tagout devices, and the responsibility for them;

4.   Specific requirements for testing a machine or equipment to determine and verify the effectiveness of lockout devices, tagout devices, and other energy control measures.

Lockout/Tagout Devices

Locks, tags, chains, wedges, key blocks, adapter pins, self-locking fasteners, or other hardware shall be provided by the employer for isolating, securing, or blocking of machines or equipment from energy sources. They shall be singularly identified; they shall be the only device(s) used for controlling energy; and they shall not be used for other purposes.

These devices must be durable and capable of withstanding the environment to which they are exposed for the maximum period of time that exposure is expected. Tagout devices shall be constructed and printed so that exposure to weather conditions, or wet and damp locations, will not cause the tag to deteriorate, or the message on the tag to become illegible. Tags and locking devices must not deteriorate when used in corrosive environments, such as areas where acid and alkali chemicals are handled and stored.

Lockout and tagout devices shall be standardized within the worksite in at least one of the following criteria: color, shape, or size, and, additionally, in the case of tagout devices, print and format shall be standardized (see Figure 69).

Lockout devices shall be substantial enough to prevent removal without the use of excessive force or unusual techniques, such as with the use of bolt cutters or other metal cutting tools. Tagout devices, including their means of attachment, shall be substantial enough to prevent inadvertent or accidental removal. Tagout device attachment means shall be of a non-reusable type, attachable by hand, self-locking, and non-releasable, with a minimum unlocking strength of no less than 50 pounds and having the general design and basic characteristics of being at least equivalent to a one-piece, all environment-tolerant nylon cable tie.

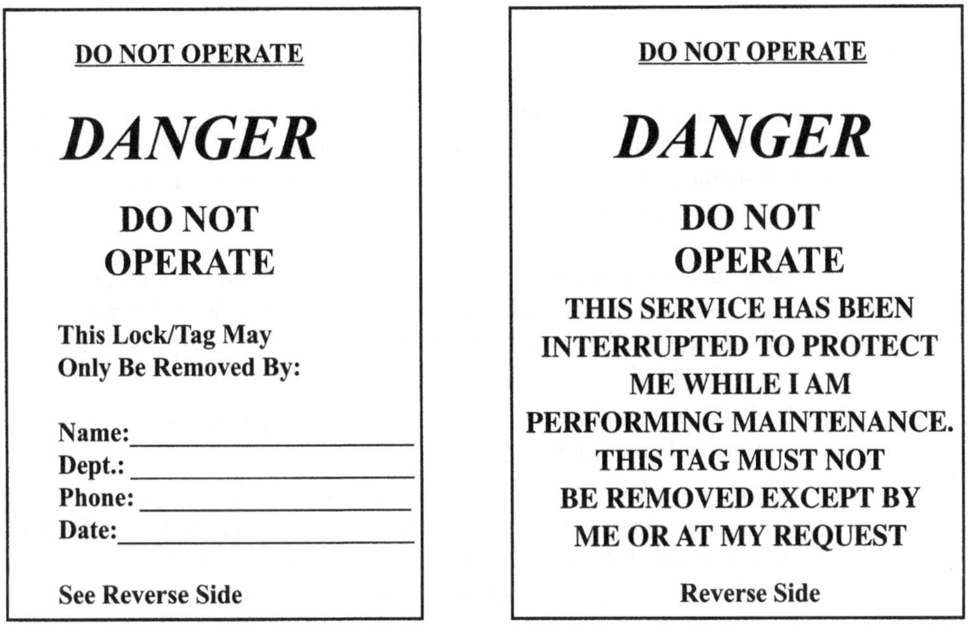

```
┌─────────────────────────┐   ┌─────────────────────────┐
│    DO NOT OPERATE       │   │    DO NOT OPERATE       │
│                         │   │                         │
│     DANGER              │   │     DANGER              │
│                         │   │                         │
│     DO NOT              │   │     DO NOT              │
│     OPERATE             │   │     OPERATE             │
│                         │   │  THIS SERVICE HAS BEEN  │
│  This Lock/Tag May      │   │ INTERRUPTED TO PROTECT  │
│  Only Be Removed By:    │   │    ME WHILE I AM        │
│                         │   │ PERFORMING MAINTENANCE. │
│  Name:_____   │   │  THIS TAG MUST NOT      │
│  Dept.: _____   │   │  BE REMOVED EXCEPT BY   │
│  Phone: _____   │   │  ME OR AT MY REQUEST    │
│  Date:_____   │   │                         │
│                         │   │                         │
│  See Reverse Side       │   │    Reverse Side         │
└─────────────────────────┘   └─────────────────────────┘
```

Figure 69. Example of a tag for use in lockout/tagout

Lockout and tagout devices shall indicate the identity of the employee applying the device(s). Tagout devices shall warn against hazardous conditions, if the machine or equipment is energized, and must include a line such as the following: "Do Not Start, Do Not Open, Do Not Close, Do Not Energize, Do Not Operate."

Periodic Inspections

The contractor should conduct a periodic inspection of the energy control procedure, at least annually, to ensure that the procedure and the requirements of this standard are being followed. The periodic inspection needs to be performed by an authorized (competent) employee, utilizing the energy control procedure being inspected. The periodic inspection should be conducted to correct any deviations or inadequacies identified. Where lockout is used for energy control, the periodic inspection shall include a review, between the inspector and each authorized employee, of that employee's responsibilities under the energy control procedure being inspected. Authorized employees are workers who lockout or tagout machines or equipment in order to perform servicing or maintenance on that machine or equipment. An affected employee becomes an authorized employee when that employee's duties include performing servicing or maintenance covered under this section.

Where tagout is used for energy control, the periodic inspection shall include a review, between the inspector and each authorized and affected employee, of that employee's responsibilities under the energy control procedure being inspected. Affected employees are those workers whose job requires him/her to operate or use a machine or equipment on which servicing or maintenance is being performed under lockout or tagout, or whose job requires him/her to work in an area in which such servicing or maintenance is being performed.

The employer should certify that the periodic inspections have been performed. The certification needs to identify the machine or equipment on which the energy control procedure was being utilized; the date of the inspection; the employees included in the inspection; and the person performing the inspection.

Training and Communications

The employer must provide training to ensure that the purpose and function of the energy control program is understood by employees, and that the knowledge and skills required for the safe application, usage, and removal of the energy controls are acquired by employees. The training shall include the following:

1. Each authorized employee shall receive training in the recognition of applicable hazardous energy sources, the type and magnitude of the energy available in the workplace, and the methods and means necessary for energy isolation and control.

2. Each affected employee shall be instructed in the purpose and use of the energy control procedure.

3. All other employees, whose work operations are or may be in an area where energy control procedures may be utilized, shall be instructed about the procedure, and about the prohibition relating to attempts to restart or reenergize machines or equipment which are locked out or tagged out.

When tagout systems are used, employees shall also be trained in the following limitations of those tags: tags are essentially warning devices affixed to energy isolating devices and they do not provide the physical restraint that is provided by a lock; they are attached to an energy isolating means and are not to be removed without authorization of the authorized person responsible for it; they are never to be bypassed, ignored, or otherwise defeated; they must be legible and understandable by all authorized employees, affected employees, and all other employees whose work operations are, or may be, in the area; in order for tags to be effective, they must be made of materials which will withstand the environmental conditions encountered in the workplace, including their means of attachment; they may evoke a false sense of security; their meaning needs to be understood as part of the overall energy control program; and they must be securely attached to energy isolating devices so that they cannot be inadvertently or accidentally detached during use.

Retraining shall be provided for all authorized and affected employees whenever there is a change in their job assignments, or a change in machines, equipment, or processes that presents a new hazard, or when there is a change in the energy control procedures. Additional retraining shall also be conducted whenever a periodic inspection reveals, or whenever the employer has reason to believe that there are deviations from, or inadequacies in the employee's knowledge or use of the energy control procedures. The retraining shall reestablish employee proficiency and introduce new or revised control methods and procedures, as necessary. The employer shall certify that employee training has been accomplished and is being kept up to date. The certification should contain each employee's name and dates of training.

Energy Isolation

Lockout or tagout shall be performed only by the authorized employees who are performing the servicing or maintenance. Affected employees are to be notified by the employer or authorized employee of the application and removal of lockout devices or tagout devices. Notification must be given before the controls are applied, and after they are removed from the machine or equipment.

Established Procedure

The established procedures for the application of energy control (the lockout or tagout procedures) need to cover the following elements and actions, and shall be done in the following sequence:

1. Prior to the preparation for shutdown or turning off a machine or equipment, the authorized or affected employee, who will be performing this task, must have knowledge of the type and magnitude of the energy, the hazards of the energy to be controlled, and the method or means to control the energy.

2. Machine or equipment shutdown must follow the procedures established for the machine or equipment. An orderly shutdown must be utilized to avoid any additional or increased hazard(s) to employees, as a result of the equipment stoppage.

3. There shall be machine or equipment isolation of all energy isolating devices that are needed to control the energy to the machine or equipment, and they shall be physically located and operated in such a manner as to isolate the machine or equipment from the energy source(s).

Lockout or Tagout Device Application

Lockout or tagout devices are to be affixed to each energy isolating device by authorized employees. Lockout devices, where used, shall be affixed in a manner that will hold the energy isolating devices in a "safe" or "off" position. Tagout devices, where used, must be affixed in such a manner that will clearly indicate that the operation or movement of energy isolating devices from the "safe" or "off" position is prohibited. Where tagout devices are used with energy isolating devices designed with the capability of being locked, the tag attachment needs to be fastened at the same point at which the lock would have been attached. Where a tag cannot be affixed directly to the energy isolating device, the tag shall be located as close as safely possible to the device and in a position that will be immediately obvious to anyone attempting to operate the device.

Stored Energy

Following the application of lockout or tagout devices to energy isolating devices, all potentially hazardous stored or residual energy is to be relieved, disconnected, restrained, and otherwise rendered safe. If there is a possibility of reaccumulation of stored energy to a hazardous level, verification of isolation shall be continued until the servicing or maintenance is completed, or until the possibility of such accumulation no longer exists.

The verification of isolation must be done by an authorized employee prior to starting work on machines or equipment that have been locked out or tagged out; the authorized employee must verify that isolation and deenergization of the machine or equipment have been accomplished.

Release from Lockout or Tagout

Before lockout or tagout devices are removed and energy is restored to the machine or equipment, procedures are to be followed and actions taken by the authorized employee(s) to ensure the following:

1. The work area must be inspected to ensure that nonessential items have been removed and machine or equipment components are operationally intact.

2. The work area must be checked to ensure that all employees have been safely positioned or removed. Before lockout or tagout devices are removed, and before machines or equipment are energized, affected employees shall be notified that the lockout or tagout devices have been removed. After lockout or tagout devices have been removed, and before a machine or equipment is started, affected employees shall be notified that the lockout or tagout device(s) have been removed.

3. Each lockout or tagout device shall be removed from each energy isolating device by the employee who applied the device. There may be an exception to this rule when the authorized employee, who applied the lockout or tagout device, is not available to remove it; that device may be removed under the direction of the employer, provided that specific procedures and training for such removal have been developed, documented, and incorporated into the employer's energy control program.

Testing or Positioning

When testing or positioning of machines, equipment, or components in situations in which lockout or tagout devices must be temporarily removed from the energy isolating device, and the machine or equipment energized to test or position the machine, equipment, or component thereof, the following sequence of actions shall be followed:

1. Clear the machine or equipment of tools and materials.

2. Remove employees from the machine or equipment area.

3. Remove the lockout or tagout devices.

4. Energize and proceed with testing or positioning.

5. Reenergize all systems and reapply energy control measures to continue the servicing and/or maintenance.

Outside Personnel (Subcontractors, etc.)

Whenever outside personnel are to be engaged in activities covered by the scope and application of the lockout/tagout procedure, the on-site employer and the outside employer shall inform each other of their respective lockout or tagout procedures. The on-site employer shall ensure that his/her employees understand and comply with the restrictions and prohibitions of the outside employer's energy control program.

Group Lockout or Tagout

When servicing and/or maintenance is performed by a crew, craft, department, or other group, they shall utilize a procedure which affords the employees a level of protection equivalent to that provided by the implementation of a personal lockout or tagout device. Group lockout or tagout devices are to be used in accordance, but not necessarily limited to, the following specific requirements:

1. Primary responsibility is vested in an authorized employee for a set number of employees working under the protection of a group lockout or tagout device (such as an operations lock).

2. Provision for the authorized employee to ascertain the exposure status of individual group members with regard to the lockout or tagout of the machine or equipment.

3. When more than one crew, craft, department, etc. is involved, assignment of the overall job-associated lockout or tagout control should be given to an authorized employee designated to coordinate affected work forces and ensure continuity of protection.

4. Each authorized employee shall affix a personal lockout or tagout device to the group lockout device, group lockbox, or comparable mechanism when he or she begins work, and shall remove those devices when he or she stops working on the machine or equipment being serviced or maintained.

Shift or Personnel Changes

Specific procedures shall be utilized during shift or personnel changes to ensure the continuity of the lockout or tagout protection, including provision for the orderly transfer of lockout or tagout device protection between off-going and oncoming employees in order to minimize exposure to hazards from the unexpected energization or start-up of the machine or equipment, or the release of stored energy.

MARINE EQUIPMENT (1926.605)

The term "longshoring operations" means the loading, unloading, moving, or handling of construction materials, equipment and supplies, etc. into, in, on, or out of any vessel from a fixed structure or shore-to-vessel, vessel-to-shore or fixed structure or vessel-to-vessel. Ramps for access of vehicles to or between barges are to be of adequate strength, provided with side boards, well maintained, and properly secured.

Unless workers can step safely to or from the wharf, float, barge, or river towboat, either a ramp or a safe walkway must be provided. Jacob's ladders are of the double rung or flat tread type. They should be well maintained and properly secured. A Jacob's ladder either hangs without slack from its lashings, or is to be pulled up entirely.

When the upper end of the means of access rests on, or is flush with, the top of the bulwark, substantial steps which are properly secured and equipped with at least one substantial hand rail approximately 33 inches in height, must be provided between the top of the bulwark and the deck. Obstructions must never be laid on or across the gangway, and the means of access is to be adequately illuminated for its full length. Unless the structure makes it impossible, the means of access is to be so located that the load will not pass over workers.

Workers are not permitted to walk along the sides of covered lighters or barges with coamings more than 5 feet high, unless there is a 3-foot clear walkway, or a grab rail, or a taut handline is provided. Decks and other working surfaces are to be maintained in a safe condition. Workers are not permitted to pass fore and aft, over, or around, deckloads, unless there is a safe passage. Workers are not permitted to walk over deckloads from rail to coaming unless there is a safe passage. If it is necessary to stand at the outboard or inboard edge of the deckload where less than 24 inches of bulwark, rail, coaming, or other protection exists, all must be provided with a suitable means of protection against falling from the deckload.

There must be in the vicinity of each barge in use at least one U.S. Coast Guard approved 30-inch life ring with not less than 90 feet of line attached, and at least one portable or permanent ladder which will reach the top of the apron to the surface of the water. If the above equipment is not available at the pier, the employer must furnish it during the time that the worker is working the barge. Workers walking or working on the unguarded decks of barges are to be protected with U.S. Coast Guard-approved work vests or buoyant vests.

MATERIAL HANDLING AND STORAGE (1926.250)

Materials handling accounts for approximately 40 percent of the lost-time incidents that occur in the construction industry. These injuries are often a result of inadequate planning, administrative, and/or engineering approaches. Therefore, in an effort to reduce workplace injuries, the following safe work procedures will be implemented and enforced at all construction projects.

Materials that are stored in tiers must be stacked, racked, blocked, interlocked, or otherwise secured to prevent sliding, falling, or collapse. Noncompatible materials are to be segregated in storage. Bagged materials must be stacked by stepping back the layers and cross-keying the bags at least every 10 bags high. Brick stacks are not to be more than 7 feet in height. When a loose brick stack reaches a height of 4 feet, it is to be tapered back 2 inches in every foot of height above the 4-foot level. When masonry blocks are stacked higher than 6 feet, the stack is to be tapered back one-half block per tier above the 6-foot level.

Also, lumber is to be stacked on a level surface and solidly supported sills, and so stacked as to be stable and self-supporting. Lumber piles must not exceed 20 feet in height, except lumber that is to be handled manually is not to be stacked more than 16 feet high. Any used lumber must have all nails withdrawn before stacking. Structural steel, poles, pipe, bar stock, and other cylindrical materials, unless racked, are to be stacked and blocked so as to prevent spreading or tilting (see Figure 70).

Figure 70 . Securely stacked cylindrical material

Materials stored inside buildings which are under construction are not to be placed within 6 feet of any hoistway or inside floor openings, nor within 10 feet of an exterior wall which does not extend above the top of the material stored. The maximum safe load limits of floors within buildings and structures, in pounds per square foot, must be conspicuously posted in all storage areas, except for floors or slabs on grade. The maximum safe loads are not to be exceeded. Materials are not to be stored on scaffolds or runways in excess of supplies needed for immediate operations.

Also, all aisles and passageways should be kept clear to provide for the free and safe movement of material handling equipment or employees, and these are to be kept in good repair. When a difference in road or working levels exists, means, such as ramps, blocking, or grading, are to be used to ensure the safe movement of vehicles between the two levels.

Any time a worker is required to work on stored material in silos, hoppers, tanks, and similar storage areas, they are to be equipped with personal fall arrest equipment which meets the requirements of Subpart M.

"Housekeeping" storage areas are to be kept free from accumulation of materials that constitute hazards from tripping, fire, explosion, or pest harborage. Vegetation control is to be exercised when necessary.

When portable and powered dockboards are used, they should be strong enough to carry the load imposed on them. Portable dockboards are to be secured in position, either by being anchored or equipped with devices which will prevent their slipping. Handholds, or other effective means, are to be provided on portable dockboards to permit safe handling. Positive protection must be provided to prevent railroad cars from being moved while dockboards or bridge plates are in position.

When handling materials, do not attempt to lift or move a load that is too heavy for one person–get help. Attach handles or holders to the load to reduce the possibility of pinching or smashing fingers.

Wear protective gloves and clothing (i.e., aprons), if necessary, when handling loads with sharp or rough edges. When pulling or prying objects, be sure you are properly positioned.

During the weekly "tool-box" meetings, employees should receive instructions on proper materials handling practices so that they are aware of the following types of injuries associated with manual handling of materials:

1. Strains and sprains from lifting loads improperly, or from carrying loads that are too heavy or large.

2. Fractures and bruises caused by dropping or flying materials, or getting hands caught in pinch points.

3. Cuts and abrasions caused by falling materials which have been improperly stored, or by cutting securing devices incorrectly.

Engineering controls should be used, if feasible, to redesign the job so that the lifting task becomes less hazardous. This includes reducing the size or weight of the object lifted, changing the height of a pallet or shelf, or installing a mechanical lifting aid.

MATERIAL HANDLING EQUIPMENT (1926.602)

Material handling equipment must include the following types of earthmoving equipment: scrapers, loaders, crawlers or wheel tractors, bulldozers, off-highway trucks, graders, agricultural and industrial tractors, and similar equipment (see Figure 71).

Seat belts are to be provided on all equipment covered by this section, and meet the requirements of the Society of Automotive Engineers, J386-1969, Seat Belts for Construction Equipment. Seat belts for agricultural and light industrial tractors must meet the seat belt requirements of Society of Automotive Engineers J333a-1970, Operator Protection for Agricultural and Light Industrial Tractors. Seat belts are not needed for equipment which is designed only for standup operation. Seat belts are not needed for equipment which does not have roll-over protective structure (ROPS) or adequate canopy protection.

No construction equipment or vehicles are to be moved upon any access roadway or grade unless the access roadway or grade is constructed and maintained to safely accommodate

Figure 71. Example of material handling equipment

the movement of the equipment and vehicles involved. Every emergency access ramp and berm used by an employer is to be constructed to restrain and control runaway vehicles. All earth-moving equipment is to have a service braking system capable of stopping and holding the equipment fully loaded, as specified in Society of Automotive Engineers SAE-J237, Loader Dozer-1971, J236, Graders-1971, and J319b, Scrapers-1971. Brake systems for self-propelled, rubber-tired, off-highway equipment, manufactured after January 1, 1972, must meet the applicable minimum performance criteria set forth in the following Society of Automotive Engineers Recommended Practices:

> Self-Propelled Scrapers .... SAE J319b-1971.
>
> Self-Propelled Graders .... SAE J236-1971.
>
> Trucks and Wagons .... SAE J166-1971.
>
> Front-End Loaders and Dozers .... SAE J237-1971.

Pneumatic-tired, earth-moving haulage equipment (trucks, scrapers, tractors, and trailing units), whose maximum speed exceeds 15 miles per hour, is to be equipped with fenders on all wheels, unless the employer can demonstrate that uncovered wheels present no hazard to personnel from flying materials. Rollover protective structures should be found on all material handling equipment (see Figure 72).

Figure 72. A scraper with rollover protection

All bidirectional machines, such as rollers, compactors, front-end loaders, bulldozers, and similar equipment, are to be equipped with a horn, distinguishable from the surrounding noise level, and they are to be operated, as needed, when the machine is moving in either direction. The horn shall be maintained in an operative condition. No earth-moving or compacting equipment, which has an obstructed view to the rear, is permitted to be used in the reverse gear unless the equipment has, in operation, a reverse signal alarm which is distinguishable from the surrounding noise level, or there is an authorized employee who signals that it is safe to do so (see Figure 73.)

Figure 73. Front-end loaders are bidirectional and required to have a back-up alarm

Scissor points on all front-end loaders, which constitute a hazard to the operator during normal operation, are to be guarded.

Tractors must have seat belts. They are required for the operators when they are seated in the normal seating arrangement for tractor operation, even though back-hoes, breakers, or other similar attachments are used on these machines for excavating or other work which result in other than normal seating.

Industrial trucks such as lift trucks, forklifts, stackers, etc. must have the rated capacity clearly posted on the vehicle so as to be clearly visible to the operator. When auxiliary removable counterweights are provided by the manufacturer, corresponding alternate rated capacities also must be clearly shown on the vehicle. These ratings shall not be exceeded. No modifications or additions which affect the capacity or safe operation of the equipment shall be made without the manufacturer's written approval. If such modifications or changes are made, the capacity, operation, and maintenance instruction plates, tags, or decals must be changed accordingly. In no case shall the original safety factor of the equipment be reduced. If a load is lifted by two or more trucks working in unison, the proportion of the total load carried by any one truck must not exceed its capacity. Steering or spinner knobs are not to be attached to the steering wheel unless the steering mechanism is of a type that prevents road reactions from causing the steering handwheel to spin. The steering knob shall be mounted within the periphery of the wheel.

All high lift rider industrial trucks are to be equipped with overhead guards which meet the configuration and structural requirements as defined in paragraph 421 of American National Standards Institute B56.1-1969, Safety Standards for Powered Industrial Trucks. All industrial trucks in use shall follow the applicable requirements of design, construction, stability, inspection, testing, maintenance, and operation, as defined in American National Standards Institute B56.1-1969.

Unauthorized personnel are not permitted to ride on powered industrial trucks. A safe place to ride is to be provided when riding on trucks is authorized. Whenever a truck is equipped with vertical only, or vertical and horizontal controls elevatable with the lifting, carriage, or forks for lifting personnel, the following additional precautions are to be taken for the protection of the personnel being elevated. They are to use a safety platform which is firmly secured to the lifting carriage and/or forks, and a means must be provided whereby personnel on the platform can shut off power to the truck. Personnel are also to be protected from falling objects, if falling objects could occur due to the operating conditions.

MEDICAL SERVICES AND FIRST AID (CFR 1926.23 AND .50)

First aid services and provisions for medical care are to be made available, by the employer, for every employee. The employer must insure the availability of medical personnel for advice and consultation on matters of occupational health. In case of serious injury, provisions are to be made for prompt medical attention prior to the commencement of the project. When there is no infirmary, clinic, physician, or hospital that is reasonably located, in terms of time and distance to the worksite, for the treatment of injured employees, a person who has a valid certificate in first aid training from the U.S. Bureau of Mines, the American Red Cross, or equivalent training, and can be verified by documentary evidence, this employee shall be available at the worksite to render first aid. First-aid supplies must be approved by the consulting physician and are to be easily accessible, when required. The first aid kit must consist of materials approved by the consulting physician and must be in a weatherproof container with individual sealed packages for each type of item. The contents of the first aid kit must be checked before being sent out on each job, and must be checked, at least weekly, on each job to ensure that the expended items are replaced. Proper equipment for prompt transportation of the injured person to a physician or hospital, or a communication system for contacting necessary ambulance service, is to be provided. The telephone numbers of the physicians, hospitals, or ambulances are to be conspicuously posted. Where the eyes or body of any person may be exposed to injurious corrosive materials, suitable facilities for quick drenching or flushing of the eyes and body are to be provided within the work area for immediate emergency use.

MOTOR VEHICLES AND MECHANIZED EQUIPMENT (1926.601)

Motor vehicles that operate within an off-highway jobsite, which is not open to public traffic, must have a service brake system, an emergency brake system, and a parking brake system. These systems may use common components, and must be maintained in operable condition. Also, each vehicle must have the appropriate number of seats for occupants, and the seat belts must be properly installed.

Whenever visibility conditions warrant additional light, all vehicles, or combinations of vehicles, in use, are to be equipped with at least two headlights and two taillights in operable condition. All vehicles, or combination of vehicles, must have brake lights, in operable condition, regardless of light conditions. Also, these vehicles are to be equipped with an adequate audible warning device at the operator's station and be in operable condition.

No worker is to use motor vehicle equipment having an obstructed view to the rear, unless the vehicle has a reverse signal alarm, audible above the surrounding noise level, or the vehicle is only backed up when an observer signals that it is safe to do so.

All vehicles with cabs are to be equipped with windshields and powered wipers. Cracked and broken glass is to be replaced. Vehicles operating in areas, or under conditions that cause fogging or frosting of the windshields, are to be equipped with operable defogging or defrosting devices.

All haulage vehicles, whose pay load is loaded by means of cranes, power shovels, loaders, or similar equipment, must have a cab shield and/or canopy which is adequate to protect the operator from shifting or falling materials.

Tools and material are to be secured in order to prevent movement when transported in the same compartment with employees. Vehicles used to transport employees must have seats firmly secured and adequate for the number of employees to be carried and must also include adequate seat belts.

Trucks with dump bodies are to be equipped with positive means of support, permanently attached, and capable of being locked in position to prevent accidental lowering of the body while maintenance or inspection work is being done. Operating levers, controlling hoisting or dumping devices on haulage bodies, are to be equipped with a latch or other device which will prevent accidental starting or tripping of the mechanism. Trip handles for tailgates of dump trucks are to be so arranged that, in dumping, the operator will be in the clear. All rubber-tired motor vehicle equipment manufactured on or after May 1, 1972 is to be equipped with fenders. Mud flaps are to be used in lieu of fenders whenever motor vehicle equipment is not designed for fenders.

All vehicles in use are to be checked at the beginning of each shift to assure that the following parts, equipment, and accessories are in safe operating condition and free of apparent damage that could cause failure while in use: service brakes, including trailer brake connections; parking system (hand brake); emergency stopping system (brakes); tires; horn; steering mechanism; coupling devices; seat belts; operating controls; and safety devices. All defects are to be corrected before the vehicle is placed in service. These requirements also apply to equipment such as lights, reflectors, windshield wipers, defrosters, fire extinguishers, etc., where such equipment is necessary.

All equipment left unattended at night, adjacent to highways or construction areas, must have lights, reflectors, and/or barricades to identify location of the equipment. Supervisory personnel must ensure that all machinery and equipment is inspected prior to each use to verify that it is in safe operating condition. Rated load capacities and recommended rules of operation must be conspicuously posted on all equipment at the operator's station. An accessible fire extinguisher of 5 BC rating or higher must be available at all operator stations. When vehicles or mobile equipment are stopped or parked, the parking brake must be set. Equipment, on inclines, must have wheels chocked, as well as the parking brake set.

NONPOTABLE WATER (1926.51)

Nonpotable water is to be used for industrial or firefighting use only and must be identified by a label. It is not to be used for drinking, washing, or cooking purposes.

PERSONAL PROTECTIVE EQUIPMENT (PPE) (1926.95)

Personal protective equipment (PPE) includes protective equipment for the eyes, face, head, and extremities. This protective equipment may be such items as protective clothing, respiratory devices, protective shields, and barriers. When this equipment is needed because of hazardous processes, environmental hazards, chemical hazards, radiological hazards, or mechanical irritants which may cause injury or impairment to any part of the worker's body, by either absorption, inhalation, or physical contact, personal protective equipment is to be provided, used, and maintained in a sanitary and reliable condition. Where employees provide their own protective equipment, the employer is responsible to assure its adequacy, including proper maintenance and sanitation of such equipment.

All personal protective equipment must be of safe design and construction for the work to be performed. See Table 13 for some of the most common guidelines for the wearing of PPE.

Table 13

Common Guidelines for Wearing PPE

- Wear all required PPE for the job or task.
- Inspect all PPE for wear or damage prior to use.
- Take care of and clean PPE when necessary.
- Do not use PPE for which worker has not received training.
- Workers working in areas where overhead structures, equipment, or stored materials create a hazard shall wear hard hats and be required to wear them at all times.
- Workers wearing prescription eye glasses should use hardened safety glass lenses.
- Goggles shall be worn during work where flying particles are definite eye hazards.
- Hand protection is required during climbing, lifting, or potential contact with chemicals.
- Approved ear protection shall be worn when required.
- Respirators shall be worn if the concentration of dust, toxic fumes, or other air contaminants exceeds safe exposure levels.
- Safety harnesses and life lines shall be used when working surfaces are above 6 feet.
- Proper work shoes or boots, in good repair, shall be worn on the jobsites.
- Workers shall wear suitable work clothing consisting of at least long pants and a tucked-in short sleeve shirt.
- Rubber boots must be worn when doing concrete work.
- Always wear life jackets when working over, or adjacent to deep water.

PILE DRIVING (1926.603)

Pile driving requires special equipment and unique skills compared with operating other construction equipment and operations. In pile driving there is a need for tremendous amounts of energy. [This energy is in the form of steam in a pressurized vessel which presents a different type of hazard to construction operations] (see Figure 74).

Figure 74. Example of piles being driven

All pressure vessels which are a part of, or used with pile driving equipment must meet the applicable requirements of the American Society of Mechanical Engineers, Pressure Vessels (Section VIII). Boilers and piping systems which are also part of, or used with pile driving must meet the applicable requirements of the American Society of Mechanical Engineers, Power Boilers (Section I).

Workers are to be provided overhead protection which does not obscure the vision of the operator and which meets the requirements of Subpart N of this part. The protection shall be the equivalent of 2-inch planking or other solid material of equivalent strength.

A stop block is also to be provided for the leads in order to prevent the hammer from being raised against the head block. While employees are working under the hammer, a blocking device, capable of safely supporting the weight of the hammer, is to be provided.

Guards are to be provided across the top of the head block to prevent the cable from jumping out of the sheaves. And, when the leads must be inclined in the driving of batter piles, provisions are to be made to stabilize the leads.

Fixed leads are to be provided with a ladder and adequate rings, or similar attachment points, so that the loft worker may engage his safety belt lanyard to the leads. If the leads are provided with loft platforms(s), such platform(s) shall be protected by standard guardrails.

A steam hose, which leads to a steam hammer or jet pipe, is to be securely attached to the hammer with an adequate length of at least 1/4-inch diameter chain or cable in order to prevent whipping in the event the joint at the hammer is broken. Air hammer hoses are to be provided with the same protection as required for steam lines. Safety chains, or equivalent means, are to be provided for each hose connection to prevent the line from thrashing around in case the coupling becomes disconnected. Steam line controls must consist of two shutoff valves, one of which is a quick-acting lever type which is within easy reach of the hammer operator.

Guys, outriggers, thrustouts, or counterbalances are to be provided, as necessary, to maintain stability of pile driver rigs. Barges or floats supporting pile driving operations must meet the applicable requirements of marine operations.

In order to protect the workers, engineers and winchmen are to only accept signals from the designated signalmen. All employees are to be kept clear when piling is being hoisted into the leads. When piles are being driven in an excavated pit, the walls of the pit are to be sloped to the angle of repose, or sheet-piled and braced. When steel tube piles are being "blown out," employees are to be kept well beyond the range of falling materials. When it is necessary to cut off the tops of driven piles, pile driving operations are to be suspended, except where the cutting operations are located, at least twice the length of the longest pile from the driver. When driving jacked piles, all access pits are to be provided with ladders and bulkheaded curbs to prevent material from falling into the pit.

PNEUMATIC TOOLS (1926.302)

Pneumatic tools are powered by compressed air and include chippers, drills, hammers, and sanders (see Figure 75). Only assigned, qualified operators should operate power, powder-actuated, or air driven tools. The following safe work procedures should be implemented and enforced at all company construction projects.

Figure 75. Example of pneumatic driven grinder

1.   Pneumatic tools that shoot nails, rivets, or staples, and operate at pressures more than 100 pounds per square inch, must be equipped with a special device to keep fasteners from being ejected unless the muzzle is pressed against the work surface.

2.   Eye protection is required and face protection recommended for employees working with pneumatic tools.

3.   Compressed air guns should never be pointed toward anyone.

4.   A safety clip or retainer must be installed to prevent attachments from being unintentionally shot from the barrel of the tool.

5.   When using pneumatic tools, check to see that they are fastened securely to the hose to prevent them from becoming disconnected. All hoses exceeding 1/2 inch inside diameter must have a safety device at the supply source, or a branch line to reduce pressure in the event of hose failure.

6.   Workers operating a jackhammer are required to wear safety glasses, shoes, and hearing protection.

Compressed air is not to be used for cleaning purposes, except where reduced to less than 30 psi, and then only with effective chip guarding and personal protective equipment.

The 30 psi requirement does not apply for concrete form, mill scale, and similar cleaning purposes.

The manufacturer's safe operating pressure for hoses, pipes, valves, filters, and other fittings is not to be exceeded. The use of hoses for hoisting or lowering tools is not permitted. All hoses exceeding 1/2-inch inside diameter must have a safety device at the source of supply, or branch line to reduce pressure in case of hose failure.

Airless spray guns, which atomize paints and fluids at high pressures (1,000 pounds or more per square inch), are to be equipped with an automatic or visible manual safety devices which prevents pulling the trigger and releasing any paint or fluid until the safety device is manually released.

In lieu of the above, a diffuser nut may be provided which prevents high pressure, high velocity release, when the nozzle tip is removed, plus a nozzle tip guard which prevents the tip from coming into contact with the operator, or other equivalent protection.

POTABLE WATER (1926.51)

An adequate supply of drinking (potable) water, which meets USPHS standard 40 CFR 72, is to be available at each place of employment. Potable water containers are to be clearly marked as drinking water, be fitted with a tap, and be capable of being tightly closed. No dipping from a drinking water container is allowed, nor can a common cup be used. If single service cups are used, a sanitary container for unused cups is to be provided, as well as a receptacle for disposing of the used cups (see Figure 76).

Figure 76. Potable water container with disposable cups

POWDER-ACTUATED GUNS (1926.302)

Powder-actuated tools operate like a loaded gun and should be treated with the same respect and precautions. Only assigned, qualified operators shall operate power, powder-actuated, or air-driven tools (see Figure 77). The following safe work procedures must be implemented and enforced at all company construction projects.

1.   All powder-actuated tools must meet American National Standards Institute, A10.3-1970, Safety Requirements for Explosive-Actuated Fastening Tools requirements for design, operation, and maintenance.

2. Never use powder-actuated tools in an explosive or flammable atmosphere.

3. Before using a powder-actuated tool, the worker should inspect it to determine that it is clean, that all moving parts operate freely, and that the barrel is free from obstructions.

4. Never point the tool at anyone, or carry a loaded tool from one job to another.

5. Do not load a tool unless it is to be used immediately. Never leave a loaded tool unattended, especially where it would be available to unauthorized persons.

6. Suitable eye and face protection are essential when using a powder-actuated tool.

7. In case of misfire, the operator should hold the tool in the operating position for at least 30 seconds, and then attempt to operate the tool for a second time. If the tool misfires again, wait another 30 seconds (still holding the tool in the operating position) and then proceed to remove the explosive load from the tool, in strict accordance with the manufacturer's instructions.

8. If the tool develops a defect during use, it should be tagged and taken out of service immediately until it is properly repaired.

9. Warning signs should be posted within the area of operation of any powder-actuated tool.

10. Powder-actuated tool operators must be qualified and carry a card certifying this fact at all times. Failure to comply with any or all safety procedures governing the use of powder-actuated tools will be sufficient cause for the immediate revocation of the operator's card.

11. Always store powder-actuated tools in a container by themselves.

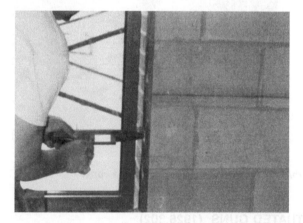

Figure 77. Powder-actuated tool being used to drive fasteners into masonry wall

The powder-actuated tool is to be tested each day before loading to see that safety devices are in proper working condition. The method of testing shall be in accordance with the manufacturer's recommended procedure. Fasteners are not to be driven into very hard or brittle materials including, but not limited to cast iron, glazed tile, surface-hardened steel, glass block, live rock, face brick, or hollow tile. Driving into easily penetrated materials is to be avoided unless such materials are backed by a substance that will prevent the pin or fastener from passing completely through and creating a flying missile hazard on the other side. And no fastener is to be driven into a spalled area caused by an unsatisfactory fastening.

POWER TOOLS (1926.300)

The following general guidelines apply to those workers who operate power tools. These workers should remove from service any damaged or worn power tools. They should never use power tools with missing guards, always wear safety shoes, gloves, and safety glasses, and never use electric power outdoors or in wet conditions without a GFCI.

All hand held power tools must be equipped with a deadman switch, and a positive "on-off" control must be provided on all hand-held powered platen sanders, grinders with wheels 2-inch diameter or less, and routers, planers, laminate trimmers, nibblers, shears, scroll saws, and jigsaws with blade shanks one-forth of an inch wide or less. A momentary contact "on-off" control must be provided on all hand-held powered drills, tapers, fasteners drivers, horizontal, vertical, and angle grinders with wheels greater than 2 inches in diameter. Tools designed to accommodate guards must be equipped with such guards when in use. All rotating, reciprocating, or moving parts of equipment (belts, gears, shafts, flyheads, etc.) must be guarded to prevent contact by employees using such equipment. Guarding must meet requirements set forth in ANSI B15.1-1953. All hand-held power tools (e.g., circular saws, chain saws, and percussion tools) without a positive accessory holding means must be equipped with a constant pressure switch that will shut off the power when pressure is released (see Figure 78).

Figure 78.   Circular saw with blade guard and equipped with a constant pressure switch

POWER TOOLS, ELECTRICAL (1926.302)

All electrical power operated tools are to be either of the approved double-insulated type or grounded in accordance with Subpart K. The use of electric cords for hoisting or lowering tools is not permitted.

Electric tools present several dangers to the user; the most serious is the possibility of electrocution (see Figure 79). Only assigned, qualified operators shall operate power, powder-actuated, or air driven tools. The following safe work procedures should be implemented and enforced at all company construction projects.

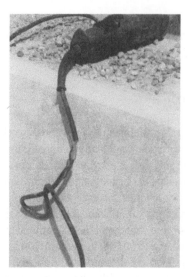

Figure 79. Damaged electrical power tool on a construction site which should be removed from service

1. Tools must have either a three-wire cord with ground and be grounded, double insulated, or powered by a low-voltage isolation transformer. A Ground-Fault Circuit Interrupter (GFCI) must be used, or the tool must be double-insulated to prevent the worker from electrical shock hazards.

2. Never remove the third prong from the plug.

3. Electric tools should be operated within their design limitations.

4. Gloves and safety footwear are recommended during use of electric tools.

5. When not in use, tools should be stored in a dry place.

6. Electric tools should not be used in damp or wet locations.

7. Work areas should be well lighted.

POWER TOOLS, FUEL DRIVEN (1926.302)

All fuel powered tools are to be stopped while being refueled, serviced, or maintained, and fuel is to be transported, handled with care, and stored to prevent fires. When fuel powered tools are used in enclosed spaces, the applicable requirements for concentrations of toxic gases and the use of personal protective equipment apply.

POWER TOOLS, HYDRAULIC (1926.302)

The fluid used in hydraulic powered tools is to be fire-resistant fluids approved under Schedule 30 of the U.S. Bureau of Mines, Department of the Interior, and must retain its operating characteristics at the most extreme temperatures to which it will be exposed. The manufacturer's safe operating pressures for hoses, valves, pipes, filters, and other fittings are not to be exceeded.

POWER TRANSMISSION AND DISTRIBUTION (1926.950)

Before beginning power transmission and distribution work, all potential hazardous conditions are to be determined by an inspection or a test. Such conditions include, but are not limited to energized lines and equipment, including power and communications, cable television and fire alarm circuits. All electrical equipment or lines are considered to be energized until determined otherwise by testing, or until grounded. It is important to determine the operating voltage of equipment and lines before working on or near energized parts. Crews working on lines or equipment must assure that the means of disconnect are open or locked out. Workers who work on energized lines are to be trained in emergency and first aid procedures. Workers working near or over water are to be protected from drowning. During night operations, spot lights or portable lights are to be used.

Tools and Protective Equipment (1926.951)

In no case are conductive objects without insulated handles allowed closer than the specifications stated in Table 14. All rubber protective equipment must comply with the provisions of ANSI J6-1950 (R-1971) series and must be visually inspected before use and rubber gloves are to be "air tested." Metal or conductive ladders are never to be used near energized lines. All tools are to be inspected, and the defective ones removed from service. Portable electric hand tools are to be double insulated. No metal measuring tapes or ropes are to be used near energized parts. Hydraulic and pneumatic tools which are used around energized lines must have nonconductive hoses. Hydraulic fluids used for insulating sections on equipment must be the insulating type.

Table 14

Alternating Current – Minimum Distances
Courtesy of OSHA

Voltage range (phase to phase) (kilovolt)	Minimum working and clear hot stick distance
2.1 to 15	2 ft. 0 in.
15.1 to 35	2 ft. 4 in.
35.1 to 46	2 ft. 6 in.
46.1 to 72.5	3 ft. 0 in.
72.6 to 121	3 ft. 4 in.
138 to 145	3 ft. 6 in.
161 to 169	3 ft. 8 in.
230 to 242	5 ft. 0 in.
345 to 362	(1)7 ft. 0 in.
500 to 552	(1)11 ft. 0 in.
700 to 765	(1)15 ft. 0 in.

NOTE: For 345-362 kv., 500-552 kv., and 700-765 kv., minimum clear hot stick distance may be reduced provided that such distances are not less than the shortest distance between the energized part and the grounded surface.

Mechanical Equipment (1926.952)

All mechanical equipment must be visually inspected each time it is used. Aerial-lift trucks are to be grounded or barricaded since they are considered energized equipment. Except for equipment certified to work on specific voltages, mechanical equipment is not to be operated closer than the distances specified in Table 15. Workers are not allowed to pass materials from an aerial-lift to a utility pole or structure unless conductor are protected (insulated).

Table 15

Minimum Clearance Distances for Live-Line, Bare-Hand Work (Alternating Current) Courtesy of OSHA

Voltage range (phase to phase) kilovolts	Distance in feet and inches for maximum voltage	
	Phase to ground	Phase to phase
2.1 to 15	2'0"	2'0"
15.1 to 35	2'4"	2'4"
35.1 to 46	2'6"	2'6"
46.1 to 72.5	3'0"	3'0"
72.6 to 121	3'4"	4'6"
138 to 145	3'6"	5'0"
161 to 169	3'8"	5'6"
230 to 242	5'0"	8'4"
345 to 362	(1)7'0"	(1)13'4"
500 to 552	(1)11'0"	(1)20'0"
700 to 765	(1)15'0"	(1)31'0"

For 345-362kv., 500-552kv., and 700-765kv., the minimum clearance distance may be reduced provided the distances are not made less than the shortest distance between the energized part and the grounded surface.

Material Handling (1926.953)

Before unloading steel poles, cross arms, or similar materials, they are to be inspected to assure they have not shifted. Poles being transported are to be secured with a red flag attached to the end of longest pole. Materials are not to be stored under energized buses or conductors. If materials are stored under conductors, then the distance found in Table 2-3 applies, and extreme caution is required. To protect workers, taglines or other suitable devices are to be used during hoisting (see Figure 80). Workers are not to work under hoisted or suspended loads unless the loads are properly supported.

Figure 80. Worker using a tagline to direct a load

Grounding for Protection of Employees (1926.954)

All conductors and equipment are to be treated as energized until deenergized or grounded. All new lines are considered deenergized when they are grounded, no induced voltage is present, or adequate clearances are maintained. All bare wire communications conductors, on poles or structures, are to be considered energized. Conductors or equipment are to be tested to assure they are deenergized prior to grounding. The ground end must always be attached first, and using insulated tools, the other end is to be attached to the conductors or equipment. Grounds must be placed at the work location, or between the work location and all sources of energy. A ground is to exist at each work location along the conductor on a line section. When making a ground is impractical or too dangerous, then work is considered energized. When grounds are being removed, the ground side is removed first, using insulated tools. During testing, grounds are removed only for testing, and with great caution. Ground electrodes must have a low resistance to ground in order to protect workers. Adequately rated and designed tower clamps are to be used to ground towers. All ground leads for tower are to be driven into the ground and capable of conducting the fault current while having a minimum conductance of a No. 2 AWG Copper.

Overhead Lines (1926.955)

Any structure which cannot support a climber must be guyed, braced, or made safe. Poles, ladders, and elevated structures are to be inspected to assure they can support a climber. When standing on the ground, workers are not to touch equipment or machinery next to energized lines or equipment unless protective equipment exists.

Prior to the removal or installing of wires or cables, the strain on poles or structures is to be determined. Poles being set, removed, or moved by using hoists or other devices are to be protected from contacting energized lines. Pole holes are not to be left unattended or unguarded. Lifting equipment is to be bonded to an effective ground and is to be considered energized when near energized equipment or lines. Taglines shall be of nonconductive type.

Metal Tower Construction

While working in unstable material during metal tower construction, the excavation for pad- or pile-type footings, which are in excess of 5 feet deep, shall be either sloped to the angle of repose, or shored, if entry is required. Ladders shall be provided for access to pad- or pile-type footing excavations in excess of 4 feet.

A designated employee shall be used in directing mobile equipment adjacent to footing excavations. No one shall be permitted to remain in the footing while equipment is being spotted for placement. When it is necessary to assure the stability of mobile equipment, the location to be used for such equipment must be graded and leveled.

When working at two or more levels on a tower assembly, employees are to have a minimum exposure to falling objects. Guy lines are to be used, as necessary, to maintain sections, or parts of sections, in position and to reduce the possibility of tipping. Members and sections being assembled shall be adequately supported. The construction of transmission towers, and the erecting of poles, hoisting machinery, site preparation machinery, and other types of construction machinery shall conform to the normal and required construction requirements.

No one shall be permitted under a tower which is in the process of erection or assembly, except as may be required to guide and secure the section being set. When erecting towers using hoisting equipment adjacent to energized transmission lines, the lines shall be deenergized, when practical. If the lines are not deenergized, extraordinary caution shall be exercised to maintain the minimum clearance distances. Erection cranes are to be set on firm level foundations, and when the cranes are so equipped, outriggers shall be used. Taglines should be utilized to maintain control of tower sections being raised and positioned, except where the use of such lines would create a greater hazard. Loadlines are not to be detached from a tower section until the section is adequately secured.

Except during emergency restoration procedures, erection shall be discontinued in the event of high wind or other adverse weather conditions which would make the work hazardous. Equipment and rigging are to be regularly inspected and maintained in safe operating condition. Adequate traffic control must be maintained when crossing highways and railways with equipment. A designated employee shall be utilized to determine that required clearance is maintained in moving equipment under or near energized lines.

Stringing and Removing Deenergized Conductors

Prior to stringing and removal of deenergized conductors, work procedures are to be reviewed, including equipment needs and precautions. When deenergized conductors can contact an energized conductor, the deenergized conductor is to be grounded and workers are to be insulated or isolated. Any existing line is to be deenergized and each end is to be ground on both sides of the crossover or, if energized, it is to be worked as such.

When crossing over an energized line over 600 volts, the ropes nets or guards structures are to be installed unless insulation or isolation of the live conductors from workers and the automatic reclosing feature of the circuit interrupter is made inoperative.

Tension reels, guard structures, tie lines, or other equivalent means are to be used when conductors are strung or removed. Conductor grips must not be used on wire rope unless designed for that purpose. The clipping crew needs to have a minimum of two structures clipped in between the crew and the conductor being sagged. The crews always work between grounds when working on bare conductors. The ground must remain in place until work is competed. Reliable communications is to be maintained between the reel tender and the pulling-rig operator. No other pulls shall take place until the previous pull is snubbed or dead-ended.

Stringing Adjacent to Energized Lines

The potential for induced voltage is to be assessed when stringing lines parallel to energized lines. When the potential for induced voltage exists, pulling and tension equipment is to be isolated, insulated, or grounded. The ground is to be installed between the tension reel and the first structure to ground each bare conductor, subconductor, or overhead ground. During a stringing operation, the conductor is to be grounded at the first tower adjacent to both the tension and pulling setup, and in increments no more than 2 miles from the ground. These grounds are to be left in place until all work is completed, then removed with a hot stick. These conductors are to be grounded at all dead ends or catch-off points. Conductors, subconductors, and ground conductors are to be bonded to the tower at an isolated tower to complete work on a transmission line.

Live-Line Bare-Hand Work

All deenergized lines are to be grounded when working a dead-end tower. When structures are worked, grounds must exist at all work locations to protect workers on conductors. Live-line and bare-hand work is to be done by trained workers. Workers must know the voltage rating of circuits, clearance to ground line, and voltage limitations of the aerial-lift equipment for live-line or bare-hand work. All equipment is to be designed, tested, and intended for live-line and bare-hand work. This type of work is to be supervised by a trained and qualified person.

The automatic reclosing feature of an interrupting device is to be made inoperative prior to working on energized lines or equipment. A conductive bucket liner, or other suitable conductive devices, is to be used for bonding the insulated aerial device to live-line or equipment. Workers are to be connected to the bucket liner with conductive shoes, leg clips, or other equivalent means. At times, electrostatic shielding or conductive clothing is to be provided. The outrigger on the aerial truck is to be extended prior to elevating the boom. The truck is to be bonded to an effective ground or barricade and is to be considered energized. The truck must be inspected prior to moving into the work position. Only clean and dry tools and equipment, which are intended for live-line and bare-hand work, are to be used.

An arm current test is to be made before the beginning of work each, and any time a higher voltage is to be worked on. The test must last at least 3 minutes on buckets which are in use, and leakage is not to be more than one micro-ampere per kilo-volt of the nominal line to line voltage.

Aerial lifts must have both upper and lower controls. The upper control must have override capabilities and be within easy reach of the worker. Lower controls are to be located at base of the boom and are not to be used without permission of the worker in the bucket. Before workers contact energized parts, the conductive bucket liner is to be bonded to the energized conductor and must not be removed until work is completed. The minimum clearance distances are to be printed on durable, nonconductive material posted inside the bucket and in view of the workers.

Underground Lines (1926.956)

To work on underground lines, workers must post warning signs when a cover of a manhole, handholes, or vault is removed. Before workers enter street openings or vaults, barriers, temporary covers, or other suitable guards are to be in place. Workers do not enter manholes or unvented vaults until forced ventilation is provided, or until the atmosphere is tested for oxygen deficiency, the presence of explosive gases, or toxic atmospheres, and found safe. A safety watch must be in the immediate vicinity to render assistance and may occasionally enter the manhole for short periods other than emergencies. A qualified worker can work alone for brief periods to inspect, perform housekeeping, or take readings.

When open flames are used in manholes, extra precautions are to be taken. When combustible gases or liquids may be present, the atmosphere is to be tested to assure safety.

Dangerous underground facilities are to be located to prevent worker exposure or damage to them prior to excavation. Underground facilities are to be protected to avoid damage when they are exposed. All cables found in an excavation are to be protected from damage. The cable to be worked on, when multiple cables exist, is to be identified electrically, unless it has unique characteristics. Cables are to be identified and verified before being cut or spliced. When buried cable or cable in manholes are worked, metallic sheath continuity is to be maintained by bonding across the opening, or by equivalent means.

Construction in Energized Substations (1926.957)

In order for work to proceed on an energized substation, it must be authorized by an authorized person. Energized facilities are to be identified prior to construction, and appropriate PPE is to be selected, as well as the safety precautions for workers denoted. Extra caution is to be taken when busbars and steel for towers and equipment are handled in the vicinity of energized facilities. At times it is may be necessary to deenergize equipment or lines to protect workers.

Barricades and barriers are to be erected to prevent accidental worker contact with energized lines or equipment. Signs indicating the hazard are to be posted near barricades or barriers.

Work on or near control panels is to be performed by designated workers. Precautions are to be taken to prevent jarring, vibration, or improper wiring causing accidental operation of relays or other protective devices.

Vehicles, gin poles, cranes, and other equipment in restricted or hazardous areas are to be controlled by a designated worker. All mobile cranes and derricks are to be effectively grounded when being moved or operated in close proximity to energized lines or equipment, or equipment worked as energized.

When substation fences are removed, a temporary fence, which affords similar protection, is to be provided when the site is unattended. Adequate interconnection between the temporary fence and permanent fence is to be maintained. All gates to unattended substations are to be locked, except when work is in progress.

External Load Helicopters (1926.958)

When using helicopters, the provisions in 29 CFR 1926.551 must be followed.

Lineman's Body Belts, Safety Straps, and Lanyards (1926.959)

Lineman's body belts, safety straps, and lanyards must meet ASTM standard B-117-64. Workers working at elevated locations are to use body belts, safety straps, and lanyards, except operations where their use creates a greater hazard, or other safeguards are used. Body belts, safety straps, and lanyards are to be inspected prior to each use. Damaged equipment is to be removed from service. Safety lines are never to be shock tested. They are used to lower workers during an emergency. The cushion support of a body belt must not contain any exposed rivets on the inside. Body belts have a maximum of four tool loops, four inches in the center of the back (from D-ring to D-ring) kept free.

PRECAST CONCRETE (1926.704)

Precast concrete construction has found its place in parking garages, bridges, and structures, and can be put together in somewhat of a jig puzzle fashion (see Figure 81). The large premade concrete parts of a structure require special handling, lifting, support, and an-

Figure 81. Example of construction of a precast parking garage

choring procedures. Precast concrete wall units, structural framing, and tilt-up wall panels are to be adequately supported to prevent overturning and collapse until permanent connections are completed. Lifting inserts which are embedded or otherwise attached to tilt-up precast concrete members must be capable of supporting at least two times the maximum intended load applied or transmitted to them. Lifting inserts which are embedded or otherwise attached to precast concrete members, other than the tilt-up members, must be capable of supporting at least four times the maximum intended load applied or transmitted to them (see Figure 82). Lifting hardware must be capable of supporting at least five times the maximum intended load applied or transmitted to the lifting hardware. No employee is permitted under precast concrete members being lifted or tilted into position, except those employees required for the erection of those members.

Figure 82. Lifting a piece of precast into place using lifting inserts

PROCESS CHEMICAL SAFETY MANAGEMENT (1926.64)

Over a number of years there has been a series of catastrophic releases of toxic, reactive, flammable, or explosive chemicals. These releases can result in toxic, fire, or explosion hazards and there is a definite need for their prevention. Regulations and guidelines were needed to address processes which involve chemicals at or above the specified threshold quantities, and processes which involve flammable liquids or gases on site and in one location in a quantity of 10,000 pounds or more, except for hydrocarbon fuels which are used solely as a fuel for workplace consumption (e.g., propane used for comfort heating, gasoline for vehicle refueling). As long as such fuels are not a part of a process which contains another highly hazardous chemical, these flammable liquids can be stored in atmospheric tanks or transferred. If they are kept below their normal boiling point without benefit of chilling or refrigeration, then these regulations are to be followed.

Retail facilities, oil or gas well drilling or servicing operations, or normally unoccupied remote facilities are not affected by this standard.

Requirements

Employers are required to develop a written plan of action regarding the implementation of employee participation. Employers must consult with employees and their representatives concerning the conduct and development of the process hazards analyses, and, also, the development of other elements of the process safety management in this standard. Employers must provide to the employees and their representatives access to the process hazard analyses and to all other information required under this standard.

The employer is to complete a compilation of the written process safety information before conducting any process hazard analysis required by the standard. The compilation of the written safety information enables the employer and the employees who are involved in operating the process to identify and understand the hazards involved with highly hazardous chemicals. This process safety information must include information pertaining to the hazards of the highly hazardous chemicals used or produced by the process, information pertaining to the technology of the process, and information pertaining to the equipment in the process.

Information pertaining to the hazards of the highly hazardous chemicals must consist of at least the following:

1. Toxicity information.

2. Permissible exposure limits.

3. Physical data.

4. Reactivity data.

5. Corrosivity data.

6. Thermal and chemical stability data.

7. Hazardous effects of inadvertent mixing of different materials that could foreseeably occur.

Also, information concerning the technology of the process includes at least the following:

1. A block flow diagram or simplified process flow diagram.

2. Process chemistry.

3. Maximum intended inventory.

4. Safe upper and lower limits for such items as temperatures, pressures, flows, or compositions.

5. An evaluation of the consequences of deviations, including those affecting the safety and health of employees.

Where the original technical information no longer exists, such information may be developed in conjunction with the process hazard analysis and it must be in sufficient detail to support the analysis. Information pertaining to the equipment in the process shall include

1. Materials of construction.

2. Piping and instrument diagrams (P&IDs).

3. Electrical classification.

4. Relief system design and design basis.

5. Ventilation system design.

6. Design codes and standards employed.

7. Material and energy balances for processes built after May 26, 1992.

8. Safety systems (e.g., interlocks, detection, or suppression systems).

The employer must document that the equipment complies with the recognized and generally accepted good engineering practices. For existing equipment designed and constructed in accordance with codes, standards, or practices that are no longer in general use, the employer must determine and document that the equipment is designed, maintained, inspected, tested, and operated in a safe manner.

The employer shall perform an initial process hazard analysis (hazard evaluation) on processes covered by this standard. The process hazard analysis is to be appropriate to the complexity of the process, and identify, evaluate, and control the hazards involved in the process. Employers shall determine and document the priority order for conducting process hazard analyses. This is to be based on a rationale which includes such considerations as extent of the process hazards, the number of potentially affected employees, the age of the process, and the operating history of the process. The process hazard analysis shall be conducted as soon as possible, but not later than May 26, 1997.

The employer must use one or more of the following methodologies that are appropriate to determine and evaluate the hazards of the process being analyzed: what-if, checklist, what-if/checklist, hazard and operability study (HAZOP), failure mode and effects analysis (FMEA), fault-tree analysis, or an appropriate equivalent methodology.

Process Hazard Analysis

The process hazard analysis must address the hazards of the process, the identification of any previous incident that had a potential for catastrophic consequences in the workplace, the engineering and administrative controls applicable to the hazards, and their interrelationships, such as the appropriate application of detection methodologies in order to provide early warning of releases (acceptable detection methods might include process monitoring and control instrumentation with alarms and detection hardware such as hydrocarbon sensors), the consequences of the failure of engineering and administrative controls, facility siting, human factors, and a qualitative evaluation of the possible safety and health effects on the employees in the workplace, because of the failure of controls.

The process hazard analysis is to be performed by a team with expertise in engineering and process operations, and the team must include at least one employee who has experience and knowledge specific to the process being evaluated. Also, one member of the team must be knowledgeable in the specific process hazard analysis methodology being used. The employer is to establish a system to promptly address the team's findings and recommendations; assure that the recommendations are resolved in a timely manner and that the resolution is documented; document what actions are to be taken; complete actions as soon as possible; develop a written schedule of when these actions are to be completed; communicate the actions to operating, maintenance, and other employees whose work assignments are in the process, and who may be affected by the recommendations or actions.

At least every five (5) years after the completion of the initial process hazard analysis, the process hazard analysis is to be updated and revalidated by a team which meets to assure that the process hazard analysis is consistent with the current process. Employers are to retain process hazards analyses and updates or revalidations for each process covered by this section, as well as the documented resolution of recommendations described for the life of the process.

Employer Responsibility

The employer is to develop and implement written operating procedures that provide clear instructions for safely conducting activities involved in each covered process. These procedures must be consistent with the process safety information, and must address at least the following steps for each operating phase, initial startup, normal operations, temporary operations, emergency shutdown, including the conditions under which emergency shutdown is required and the assignment of the shutdown responsibility to qualified operators, in order to ensure that emergency shutdown is executed in a safe and timely manner. The written operating procedures must also include the procedures for emergency operations, normal shutdown, startup following a turnaround or after an emergency shutdown, operating limits (including the consequences of deviation), the steps required to correct or avoid deviation, the safety and health considerations (such as properties of, and hazards presented by, the chemicals used in the process), the precautions necessary to prevent exposure (including engineering controls, administrative controls, and personal protective equipment), the control measures to be taken if physical contact or airborne exposure occurs, the quality control for raw materials, the control of hazardous chemical inventory levels, or any special or unique hazards, and safety systems and their functions.

Operating procedures are to be readily accessible to employees who work in, or maintain a process. The operating procedures are to be reviewed as often as necessary to assure that they reflect current operating practices, including changes that result from changes in process chemicals, technology, and equipment, and changes to facilities. The employer must certify annually that these operating procedures are current and accurate and is to develop and implement safe work practices to provide for the control of hazards during operations. These safe work practices include such things as lockout/tagout, confined space entry, opening process equipment or piping, and control over entrance into a facility by maintenance, contractor, laboratory, or other support personnel. These safe work practices apply to both employees and contractor employees.

Each employee presently involved in operating a process, and each employee before being involved in operating a newly assigned process, are to be trained in an overview of the process and in the operating procedures. The training shall include an emphasis on the specific safety and health hazards, emergency operations (including shutdown), and safe work practices applicable to the employee's job tasks. In lieu of the initial training, for those employees already involved in operating a process on May 26, 1992, an employer may certify in writing that the employee has the required knowledge, skills, and abilities to safely carry out the duties and responsibilities as specified in the operating procedures. In order to assure that the employees involved in operating a process understand and adhere to the current operating procedures of the process, refresher training is to be provided at least every three years, and more often if necessary. The employer, in consultation with the employees involved in operating the process, shall determine the appropriate frequency of the refresher training. The employer must ascertain that each employee involved in operating a process has received and understands the training. The employer must prepare a record which contains the identity of the employee, the date of training, and the means used to verify that the employee understood the training.

Contractors performing maintenance or repair, turnaround, major renovation, or specialty work on or adjacent to a covered process are impacted by this regulation. It does not apply to contractors providing incidental services which do not influence process safety, such as janitorial work, food and drink services, laundry, delivery, or other supply services. The employer, when selecting a contractor, must obtain and evaluate information regarding the contract employer's safety performance and programs. The employer must inform the contract employers of the known potential fire, explosion, or toxic release hazards related to the contractor's work and the process. The employer has to explain to contract employers the applicable provisions of the emergency action plan. As part of the responsibility of employers, they must develop and implement safe work practices to control the entrance, exit, and the presence of contract employers and their employees while they are in the covered process areas, and the employer must periodically evaluate the performance of the contract employers. Also, the employer must maintain a contract employee injury and illness log related to the contractor's work in process areas.

Contractor Responsibility

The contract employer must assure that each contract employee is trained in the work practices necessary to safely perform his/her job. He or she must also assure that each contract employee is instructed in the known potential of fire, explosion, or toxic release hazards related to his/her job and the process, and must have applicable provisions for the emergency action plan. The contract employer shall document that each contract employee has received and understood the training and must prepare a record which contains the identity of the contract employee, the date of training, and the means used to verify that the employee understood the training. The contract employer must assure that each contract employee follows the safety rules of the facility, including the safe work practices. The contract employer advises the employer of any unique hazards presented by the contract employer's work, or of any hazards found by the contract employer's work.

The Process

The employer performs a pre-startup safety review for new facilities and for modified facilities, when the modification is significant enough to require a change in the process safety information The pre-startup safety review shall confirm that prior to the introduction of highly hazardous chemicals to a process,

1. Construction and equipment are in accordance with design specifications.

2. Safety, operating, maintenance, and emergency procedures are in place and are adequate.

3. For new facilities, a process hazard analysis has been performed and recommendations have been resolved or implemented before startup; and modified facilities meet the requirements contained in management of change.

4. Training of each employee involved in operating a process has been completed.

The types of process equipment that are most applicable are pressure vessels and storage tanks, piping systems (including piping components such as valves), relief and vent systems and devices, emergency shutdown systems, controls (including monitoring devices and sensors, alarms, and interlocks), and pumps.

The employer shall establish and implement written procedures to maintain the ongoing integrity of process equipment. To assure that the employee can perform the job tasks in a safe manner, the employer shall train each employee, involved in maintaining the on-going integrity of process equipment, in an overview of that process, its hazards, and the procedures applicable to the employee's job tasks. Inspections and tests are to be performed on process equipment. Inspection and testing procedures must follow recognized and generally accepted good engineering practices. The frequency of inspections and tests of the process equipment is to be consistent with applicable manufacturers' recommendations and good engineering practices, and more frequently if determined to be necessary by prior operating experience. The employer must document each inspection and test that has been performed on process equipment. The documentation shall identify the date of the inspection or test, the name of the person who performed the inspection or test, the serial number or other identifier of the equipment on which the inspection or test was performed, a description of the inspection or test performed, and the results of the inspection or test. The employer must correct deficiencies in equipment that are outside acceptable limits before further use, or in a safe and timely manner when other necessary means are taken to assure safe operation.

New Operations

In the construction of new plants and equipment, the employer must assure that the equipment, as it is fabricated, is suitable for the process application for which it will be used. Appropriate checks and inspections are to be performed to assure that the equipment is installed properly and consistent with design specifications and the manufacturer's instructions. The employer must assure that maintenance materials, spare parts, and equipment are suitable for the process application for which they will be used.

Hot Work

The employer shall issue a hot work permit for hot work operations conducted on or near a covered process. The permit shall document that the fire prevention and protection requirements have been implemented prior to beginning the hot work operations and it must indicate the date(s) authorized for hot work and identify the object on which hot work is to be performed. The permit is to be kept on file until completion of the hot work operations.

Management Change

The employer must establish and implement written procedures to manage changes (except for "replacements in kind") to process chemicals, technology, equipment, and procedures, and changes to facilities that affect a covered process. The procedures should assure that the following considerations are addressed prior to any change: the technical basis for the proposed change, the impact of the change on safety and health, the modifications to operating procedures, the necessary time period for the change, and the authorization requirements for the proposed change.

Employees who are involved in the operating and maintenance of a process, and contract employees whose job tasks will be affected by a change in the process, operating procedures, or practices, shall be informed of, and trained in, the change prior to start-up of the process, or the affected part of the process. If a change occurs because of a change in the process safety information, operating procedures, or practices, this information is to be updated accordingly.

The employer shall investigate each incident which resulted in, or could reasonably have resulted in a catastrophic release of a highly hazardous chemical in the workplace. An incident investigation is to be initiated as promptly as possible, but not later than 48 hours following the incident. An incident investigation team is to be established and shall consist of at least one person knowledgeable in the process involved, including a contract employee, if the incident involved work of the contractor, and there must also be other persons with appropriate knowledge and experience who will thoroughly investigate and analyze the incident. A report is to be prepared at the conclusion of the investigation which includes, at a minimum, the date of the incident, the date the investigation began, a description of the incident, the factors that contributed to the incident, and any recommendations resulting from the investigation. The employer must establish a system to promptly address and resolve the incident report findings and recommendations. Resolutions and corrective actions are to be documented. The report is to be reviewed with all affected personnel whose job tasks are relevant to the incident findings, including contract employees, where applicable. All incident investigation reports are to be retained for five years.

Emergency Action Plan

The employer needs to establish and implement an emergency action plan for the entire plant. In addition, the emergency action plan must include procedures for handling small releases. Employers covered under this standard may also be subject to the hazardous waste and emergency response provisions.

Compliance Certification

To verify that the procedures and practices developed under the standard are adequate and are being followed, employers must certify that they have evaluated compliance at least every three years. The compliance audit is to be conducted by at least one person knowledgeable in the process. A report of the findings of the audit is to be developed. The employer must promptly determine and document an appropriate response to each of the findings of the compliance audit, and document that the deficiencies have been corrected. Employers need to retain the two (2) most recent compliance audit reports. They must also make all information that is necessary to achieve compliance available to those persons responsible for compiling the process safety information, to those assisting in the development of the process hazard analysis, to those responsible for developing the operating procedures, and to those involved in incident investigations, emergency planning, and response and compliance audits, without regard to possible trade secret status of such information. Nothing precludes the employer from requiring the persons to whom the information is made available to enter into confidentiality agreements where they cannot disclose information deemed as trade secrets. Employees and their designated representatives have access to trade secret information contained within the process hazard analysis and other documents required to be developed by this standard.

RADIATION, IONIZING (1926.53)

Keep clear of all radioactive materials or areas where work is being done with radioactive material. These areas must be barricaded and posted with a radiation hazard sign. A properly executed permit must be approved prior to bringing radioactive sources onto the worksite. In construction and related activities involving the use of sources of ionizing radiation, the pertinent provisions of the Nuclear Regulatory Commission Standards for Protection Against Radiation (10 CFR Part 20), relating to protection against occupational radiation exposure, shall apply. Any activity which involves the use of radioactive materials or X-rays, whether or not under license from the Nuclear Regulatory Commission, is to be performed by competent persons specially trained in the proper and safe operation of such equipment. In the case of materials used under Commission license, only persons actually licensed, or competent persons under the direction and supervision of the licensee, shall perform such work.

RADIATION, NON-IONIZING (LASERS) (1926.54)

Only qualified and trained employees are to be assigned to install, adjust, and operate laser equipment. Proof of the qualification of the laser equipment operator is to be available and in possession of the operator at all times. Laser equipment must bear a label to indicate maximum output and employees are not to be exposed to light intensities above

1.   Direct staring: 1 micro-watt per square centimeter.

2.   Incidental observing: 1 milliwatt per square centimeter;

3.   Diffused reflected light: 2 1/2 watts per square centimeter.

Laser units in operation should be set up above the heads of the employees, when possible. Employees are not to be exposed to microwave power densities in excess of 10 milliwatts per square centimeter. Employees, when working in areas in which a potential exposure to direct or reflected laser light greater than 0.005 watts (5 milliwatts) exists, are to be provided with antilaser eye protection devices. Areas in which lasers are used are to be posted with standard laser warning placards.

Beam shutters or caps are to be utilized, or the laser turned off when laser transmission is not actually required. When the laser is left unattended for a substantial period of time, such as during lunch hour, overnight, or at change of shifts, the laser is to be turned off.

Only mechanical or electronic means are to be used as a detector for guiding the internal alignment of the laser. The laser beam is never to be directed at employees (see Figure 83).

Figure 83. Construction worker using a laser for leveling operations

When it is raining or snowing, or when there is dust or fog in the air, the operation of laser systems is prohibited where practicable; in any event, employees are to be kept out of range of the area of source and target during such weather conditions.

RIGGING (1926.251)

Rigging Equipment for Material Handling (1926.251)

OSHA's most common citations related to rigging are those citations that involve no inspection of riggings for defects, no informed competent person, no tags on chain or synthetic slings, the use of defective alloy chains, the company's own manufacturing of their lifting devices, and deaths due to overloading or defective rigging.

The rigging of materials to be handled, moved, or lifted has to be done by individuals who have attained skills in the proper and safe rigging procedures. These skills include not only proper rigging procedures, but the knowledge to select the proper equipment for the task. Rigging equipment for material handling must be inspected prior to use on each shift, and as necessary during its use, to ensure that it is safe. Defective rigging equipment is to be removed from service. Rigging equipment is not to be loaded in excess of its recommended safe working load, which is provided in 29 CFR 1926.251 Tables H-1 through H-20. Rigging equipment, when not in use, is to be removed from the immediate work area so as not to present a hazard to employees.

The old rigging saying of "rig to the center of gravity" is critical; the load must be level. Fouling of a load or rigging gear increases loading and can affect load control. Working conditions such as wind, temperature, and dynamic loading are examples of unusual conditions which could affect load distribution.

The load is to be placed in a sling and connecting hardware which could be dramatically affected by the sling angle and/or all multiple legs (see Figure 84). All working load limits are to be based on inline loading. When side loading is allowed, the capacity is to be reduced by at least 75%

Figure 84. Using a multiple leg sling for lifting

Special custom design grabs, hooks, clamps, or other lifting accessories, for such units as modular panels, prefabricated structures, and similar materials, are to be marked to indicate the safe working loads, and are to be proof-tested prior to use to 125 percent of their rated load.

Each day before being used, the sling and all fastenings and attachments are to be inspected for damage or defects, by a competent person designated by the employer. Additional inspections are to be performed during sling use, where service conditions warrant. Damaged or defective slings must be immediately removed from service.

Welded Alloy Steel Chains (1926.251)

As part of the material handling and rigging process, chains are often the tool of choice (see Figure 85). As with other slings, chains should be inspected upon purchase, before each use, and annually or as dictated by use. These inspections are to be performed by the employer's competent person(s). During these inspections, the competent person should look for wear, nicks, cracks, breaks, gouges, stretch, bends, weld splatter, and temperature damage.

Figure 85. Chain using a choker hitch for lifting.

Chains should have a safety factor of four to one. All alloy steel chains should have an identification tag which is durable and permanently affixed and states the size, grade, rated capacity, and reach. Chains should never be used to lift loads which are not equal to or less than the manufacturer's rated capacities. Any hooks, rings, oblong links, pear shaped links, welded or mechanical coupling links, or other attachments are to be rated at least equal to the chain you are using. If this is not the case, then the chain's rated lifting capacity must be reduced to the weakest component.

Welded alloy steel chain slings are to have permanently affixed durable identification stating size, grade, rated capacity, and sling manufacturer. Hooks, rings, oblong links, pear-shaped links, welded or mechanical coupling links, or other attachments, when used with alloy steel chains, must have a rated capacity at least equal to that of the chain. Job or shop hooks and links, or makeshift fasteners, formed from bolts, rods, etc., or other such attachments, are never to be used.

In addition to the daily inspections, a thorough periodic inspection of alloy steel chain slings in use is to be made on a regular basis. The inspection is to be determined by the frequency of sling use, the severity of service conditions, the nature of the lifts being made, and the experience gained on the service life of slings used in similar circumstances. Such inspections are not to be performed at intervals greater than once every 12 months. The employer must make and maintain a record of the most recent month in which each alloy steel chain sling was thoroughly inspected, and make such records available for examination.

Wire Ropes (1926.251)

When wire rope is used, the safe working loads must be determined for the various sizes and classifications of the improved plow steel wire rope and wire rope slings with various types of terminals. For sizes, classifications, and grades not included in Tables H-3 through H-13 of 29 CFR 1926.251, the safe working load, which is recommended by the manufacturer for specific, identifiable products, is to be followed, provided that a safety factor of not less than 5 is maintained.

Figure 86. Example of wire rope slings being used

Protruding ends of strands in splices on slings and bridles are to be covered or blunted. Wire rope is not to be secured by knots, except on haul back lines on scrapers. An eye splice made in any wire rope is not to be less than three full tucks. However, this requirement does not operate to preclude the use of another form of splice or connection which can be shown to be as efficient and which is not otherwise prohibited.

Except for eye splices at the ends of wires, and for endless rope slings, each wire rope used in hoisting, lowering, or pulling loads, must consist of one continuous piece without a knot or splice. Eyes in wire rope bridles, slings, or bull wires are not to be formed by wire rope clips or knots.

Wire rope is not used if, in any length of eight diameters, the total number of visible broken wires exceeds 10 percent of the total number of wires, or if the rope shows other signs of excessive wear, corrosion, or defect. See Figure 87 for an example of a lay.

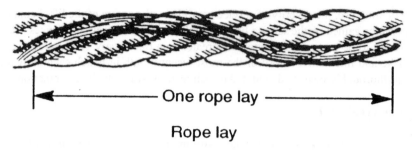

Rope lay

Figure 87. Example of rope lay. Courtesy of Department of Energy

When used for eye splices, the U-bolt is to be applied so that the "U" section is in contact with the dead end of the rope (see Figure 88).

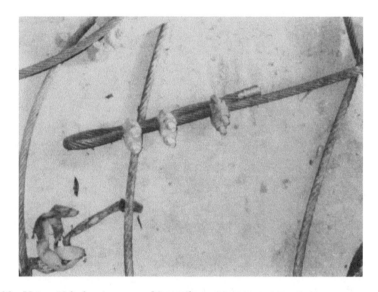

Figure 88. Using U-bolt wire rope clips to form an eye

Slings must never be shortened with knots or bolts or other makeshift devices, nor are sling legs to be kinked. Slings used in a basket hitch must have the loads balanced to prevent slippage. Slings must be padded or protected from the sharp edges of their loads. Hands or fingers shall not be placed between the sling and its load while the sling is being tightened around the load. Shock loading is prohibited. Slings are not to be pulled from under a load when the load is resting on the sling.

Cable laid and 6×19 and 6×37 slings must have a minimum clear length of wire rope 10 times the component rope diameter between splices, sleeves, or end fittings. Braided slings must have a minimum clear length of wire rope 40 times the component rope diameter between the loops or end fittings. Cable-laid grommets, strand-laid grommets, and endless slings are to have a minimum circumferential length of 96 times their body diameter.

Fiber core wire rope slings of all grades shall be permanently removed from service if they are exposed to temperatures in excess of 200 degree F. When nonfiber core wire rope slings of any grade are used at temperatures above 400 degree F, or below minus 60 degree F, recommendations by the sling manufacturer regarding use at that temperature shall be followed.

Welding of end attachments, except covers to thimbles, is to be performed prior to the assembly of the sling. All welded end attachments are not to be used unless proof tested by the manufacturer, or equivalent entity at twice their rated capacity, prior to initial use. The employer must retain a certificate of proof of test, and make it available for examination.

Synthetic Rope (1926.251)

When using natural or synthetic fiber rope slings, 29 CFR 1926.251 Tables H-15, 16, 17, and 18 apply. All splices, made to rope slings which are provided by the employer, are to be made in accordance with the manufacturer's fiber rope recommendations. In manila rope, eye splices are to contain at least three full tucks, and short splices must contain at least six full tucks (three on each side of the center line of the splice). In layed synthetic fiber rope, eye splices must contain at least four full tucks, and short splices must contain at least eight full tucks (four on each side of the center line of the splice). Strand end tails are not to be trimmed short (flush with the surface of the rope) but trimmed immediately adjacent to the full tucks. This precaution applies to both eye and short splices and all types of fiber rope. For fiber ropes under 1-inch diameter, the tails must project at least six rope diameters beyond the last full tuck. For fiber ropes 1-inch diameter and larger, the tails must project at least 6 inches beyond the last full tuck. In applications where the projecting tails may be objectionable, the tails are to be tapered and spliced into the body of the rope using at least two additional tucks (this will require a tail length of approximately six rope diameters beyond the last full tuck). For all eye splices, the eye is to be sufficiently large enough to provide an included angle of not greater than 60 degree at the splice, when the eye is placed over the load or support. Knots are never to be used in lieu of splices.

Natural and synthetic fiber rope slings, except for wet frozen slings, may be used in a temperature range from minus 20 degree F to plus 180 degree F without decreasing the working load limit. For operations outside this temperature range, and for wet frozen slings, the manufacturer's recommendations for the sling is to be followed.

Spliced fiber rope slings are not to be used unless they have been spliced in accordance with the minimum requirements, and in accordance with any additional recommendations of the manufacturer. In manila rope, eye splices must consist of at least three full tucks, and short splices must consist of at least six full tucks, three on each side of the splice center line.

For fiber rope under one inch in diameter, the tail shall project at least six rope diameters beyond the last full tuck. For fiber rope one inch in diameter and larger, the tail shall project at least six inches beyond the last full tuck. Where a projecting tail interferes with the use of the sling, the tail shall be tapered and spliced into the body of the rope, using at least two additional tucks (which will require a tail length of approximately six rope diameters beyond the last full tuck).

Fiber rope slings are to have a minimum clear length of rope between eye splices equal to 10 times the rope diameter. Knots are not to be used in lieu of splices. Clamps not designed specifically for fiber ropes are not to be used for splicing. For all eye splices, the eye is to be of such size to provide an included angle of not greater than 60 degrees at the splice, when the eye is placed over the load or support. Fiber rope slings shall not be used if end attachments in contact with the rope have sharp edges or projections.

In synthetic fiber rope, eye splices must consist of at least four full tucks, and short splices must consist of at least eight full tucks, four on each side of the center line. Strand end tails shall not be trimmed flush with the surface of the rope immediately adjacent to the full tucks. This applies to all types of fiber rope, and both eye and short splices. Natural and synthetic fiber rope slings are to be immediately removed from service if any of the following conditions are present: abnormal wear, powdered fiber between strands, broken or cut fibers, variations in the size or roundness of strands, discoloration or rotting, or distortion of hardware in the sling.

Web Slings (1926.251)

Web slings made from synthetic materials with a safety factor of 5 to 1 (ANSI B30.9A-1994) must be proof tested to twice the rated load (see Figure 89). Rated capacities are not to be exceeded. Synthetic web slings are to be permanently marked with the manufacturer's name, stock number, type of material, and rate loads for different types of hitches. As with any sling choker, hitch loads are not to exceed 80% of vertical rated load. The rated loads for non-vertical bridle or basket hitches must be adjusted in accordance with the horizontal sling angle. The horizontal angle should never be less than 30 degrees.

Figure 89. Web sling being used on construction site

Synthetic w-ebbing is to be of uniform thickness and width, and selvage edges are not to be split from the webbing's width. Fittings are to be of a minimum breaking strength equal to that of the sling and must be free of all sharp edges that could in any way damage the webbing. Stitching is to be the only method used to attach end fittings to webbing and to form eyes. The thread must be an even pattern and contain a sufficient number of stitches to develop the full breaking strength of the sling. When synthetic web slings are used, the following precautions are to be taken:

1. Nylon web slings are not to be used where fumes, vapors, sprays, mists, liquids of acids, or phenolics are present.

2. Polyester and polypropylene web slings, and slings with aluminum fittings are not to be used where fumes, vapors, sprays, mists, or liquids of caustics are present.

Web slings should be inspected when purchased, prior to each use, and annually. Although OSHA does not require recordkeeping, it is highly recommended. Inspections should be done by your competent person(s) and should be done according to the frequency of use, adverse lifting conditions, or other extenuating circumstances, such as temperature extremes.

Synthetic web slings of polyester and nylon are not to be used at temperatures in excess of 180 degree F. Polypropylene web slings are not to be used at temperatures in excess of 200 degree F. Synthetic web slings are to be immediately removed from service if any of the following conditions are present: acid or caustic burns, melting or charring of any part of the sling surface, snags, punctures, tears or cuts, broken or worn stitches, or distortion of fittings.

When inspecting web slings, look for knots/stretching, broken or worn stitches, chemical burns, cuts, snags, holes, and tears or excessive abrasions. The inspection should also include the examination of fittings for cracks, deforming, corrosion/ pitting, breakage, or missing or illegible labels. If any of these are observed, the web sling should be removed from service immediately, and destroyed or tagged out. Remember that ultraviolet or extreme sunlight can also damage web slings; this necessitates storing them out of the sunlight and in a dry area. In all cases, the manufacturer's guidelines, rate loads, and specifications should be followed.

Shackles and Hooks (1926.251)

Appendix F provides information regarding the safe working loads of various sizes of shackles. Higher safe working loads are permissible, when recommended by the manufacturer for specific, identifiable products, provided that a safety factor of not less than five is maintained, as well as hooks. The manufacturer's recommendations are to be followed in determining the safe working loads of the various sizes and types of specific and identifiable hooks. All hooks for which no applicable manufacturer's recommendations are available, are to be tested to twice the intended safe working load, before they are initially put into use (see Figure 90). The employer shall maintain a record of the dates and results of such tests.

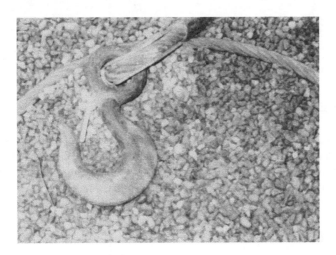

Figure 90. Example of a hook with a safety latch

ROLLOVER PROTECTIVE STRUCTURES (1926.1000)

All of the following material handling equipment, rubber-tired self-propelled scrapers, rubber-tired front-end loaders, rubber-tired dozers, wheel-type agricultural and industrial tractors, crawler-type loaders, and motor graders, must have rollover protection structures (ROPS) if manufactured after September 1, 1972. Those manufactured as far back as July 1,

1969 will need to be retrofitted.

ROPS are to be designed to support two times the weight of the equipment at the point of impact; this should protect the operator from being crushed (see Figure 91). The ROPS will not protect the operator if he or she is ejected or jumps from the operating cab of the equipment. With ROPS there should be an enforced requirement for operators to wear their seat belts. Many operators feel they can jump clear of the equipment which seldom happens because the direction and momentum of the jump is dictated largely by the direction and speed of the equipment. Suffice to say, most operators who jump are run over by their own equipment which is moving in the same direction as the jumper.

Overhead protection should be provided such that materials cannot fall through the opening in the rollover protection and strike the worker.

Figure 91. Dozer with rollover protection and overhead protection

SAFETY HARNESS

As of January 1, 1998, all workers are required to use safety harnesses, or full body harnesses for fall protection; body belts may no longer be used for fall protection. Body belts can still be used for positioning devices. These harnesses' webbing, and their accompanying hardware must posses a tensile strength of 5000 pounds (see Figure 92).

SAFETY NETS (1926.105 & 502)

Safety nets are to be used when workplaces are more than 25 feet above the surface, or over water, and when the use of ladders, scaffolds, catch platforms, temporary floors, safety lines, and safety harnesses are not practical. All safety nets are to be inspected daily, and any tools or debris must be removed from them. Safety nets are never used for falls greater than 30 feet. When safety nets are used around the perimeter of a work area, their lateral extension out from the side is to be 8 feet for a potential fall of 5 feet, 10 feet for a potential fall of greater than 5 feet, but less than 10 feet, and 13 feet for a potential fall greater then 10 feet. The maximum size of the net's mesh should be no greater than 36 square inches, or longer than 6 inches. The border ropes for the webbing must have a minimum breaking strength of 5000

Figure 92. Example of worker with a full body harness

Figure 93. Perimeter safety nets in use to protect steelworkers involved in erection

pounds. (See Figure 93 for example of safety nets.)

When installed safety nets undergo a drop test, this requires a 400 pound bag of sand, approximately 30 inches in diameter, to be dropped into the net from the highest walking/working surface, but not less than 42 inches. This should be repeated after major repairs, and every six months. If a drop test is not practical, the employer must certify and make a record attesting to the fact that it meets the compliance requirements. This activity may be done by a competent person.

SCAFFOLDS (1926.450)

The use of scaffolds, as tools for working at varied levels on construction sites, is a fixture in the construction industry (see Figure 94). Unfortunately, there have been many accidents involving scaffolding. Some scaffolding has collapsed because of improper erection, some have fallen because the scaffolding could not support the loads placed on it, and some have been improperly used because of the lack of knowledge, or the lack of proper training regarding the use of scaffolds. Scaffolds are one of the leading causes or contributors to construction fall fatalities. They must be designed by a qualified person and be constructed and loaded in accordance with that design.

Figure 94. A large scaffold system encompassing an entire building

General Requirements (1926.451)

Each scaffold and scaffold component must be capable of supporting, without failure, its own weight, and at least four times the maximum intended load applied or transmitted to it. Direct connections to the roofs and floors, and counterweights used to balance adjustable suspension scaffolds, are to be capable of resisting at least four times the tipping moment imposed by the scaffold's operating at the rated load of the hoist, or 1.5 (minimum) times the tipping moment imposed by the scaffold operating at the stall load of the hoist, whichever is greater.

Each suspension rope, including connecting hardware, used on non-adjustable or adjustable suspension scaffolds shall be capable of supporting, without failure, at least six times the maximum intended load applied or transmitted to that rope. Adjustable scaffolds must operate at either the rated load of the hoist, or two (minimum) times the stall load of the hoist, whichever is greater. The stall load of any scaffold hoist must not exceed three times its rated load.

Platforms

Each platform on all working levels of scaffolds is to be fully planked or decked between the front uprights and the guardrail supports (see Figure 95). The planking is to be installed so that the space between adjacent units, and the space between the platform and the uprights is no more than one inch wide, except where the employer can demonstrate that a wider space is necessary (for example, to fit around uprights when side brackets are used to extend the width of the platform). When the employer can demonstrate this, then the platform is planked or decked as fully as possible and the remaining open space between the platform and the uprights shall not exceed 9 1/2 inches. The requirement to provide full planking or decking does not apply to platforms used solely as walkways, or solely by employees performing scaffold erection or dismantling. In these situations, only the planking that the employer establishes as necessary to provide safe working conditions, is required. Platforms must not deflect more than 1/60 of the span when loaded. Debris is not allowed to accumulate on platforms.

Figure 95. A full planked scaffold with guardrails

Each scaffold platform and walkway needs to be at least 18 inches wide. Each ladder jack scaffold, top plate bracket scaffold, roof bracket scaffold, and pump jack scaffold is to be at least 12 inches (30 cm) wide. There is no minimum width requirement for boatswains' chairs. Where scaffolds must be used in areas that the employer can demonstrate are so narrow that platforms and walkways cannot be at least 18 inches wide, such platforms and walkways are to be as wide as feasible, and employees on those platforms and walkways shall be protected from fall hazards by the use of guardrails and/or personal fall arrest systems.

When the front edge of a platform is more than 14 inches from the face of the work, it must have a guardrail system erected along the front edge, and/or personal fall arrest systems to prevent workers from falling. The maximum distance from the face for outrigger scaffolds is 3 inches, and the maximum distance from the face for plastering and lathing operations is 18 inches.

Each end of a platform or its plank, unless cleated or otherwise restrained by hooks or equivalent means, must extend over the centerline of its support at least 6 inches. Each end of a platform or planking 10 feet or less in length is not to extend over its support more than 12 inches, and each platform greater than 10 feet in length must not extend over its support more than 18 inches, unless the platform is designed and installed so that the cantilevered portion of the platform is able to support employees and/or materials without tipping, or has guardrails which block employee access to the cantilevered end. On scaffolds where scaffold planks are abutted to create a long platform, each abutted end must rest on a separate support surface. This does not preclude the use of common support members, such as "T" sections, to support abutting planks, or hooks on platforms designed to rest on common supports. On scaffolds where platforms are overlapped to create a long platform, the overlap must occur only over supports, and is not to be less than 12 inches, unless the platforms are nailed together or otherwise restrained to prevent movement.

At all points of a scaffold where the platform changes direction, such as turning a corner, any platform that rests on a bearer at an angle other than a right angle, is laid first, and a platform which rests at a right angle over the same bearer, is laid second, on top of the first platform. Wood platforms are not to be covered with opaque finishes, except that platform edges may be covered or marked for identification. Platforms may be coated periodically with wood preservatives, fire-retardant finishes, and slip-resistant finishes; however, the coating may not obscure the top or bottom wood surfaces.

Scaffold Components

Scaffold components manufactured by different manufacturers are not to be intermixed unless the components fit together without force, and the scaffold's structural integrity is maintained by the user. Scaffold components manufactured by different manufacturers are not to be modified in order to intermix them, unless a competent person determines the resulting scaffold is structurally sound. Scaffold components made of dissimilar metals are not to be used together unless a competent person has determined that galvanic action will not reduce the strength of any component below accepted support levels.

Supporting Scaffolds

Supported scaffolds with a height to base width (including outrigger supports, if used) ratio of more than four to one (4:1) shall be restrained from tipping, by guying, tieing, bracing, or equivalent means, as follows:

1. Guys, ties, and braces are to be installed at locations where horizontal members support both inner and outer legs.

2. Guys, ties, and braces are to be installed according to the scaffold manufacturer's recommendations, or at the closest horizontal member to the 4:1 height, and to be repeated vertically at locations of horizontal members every 20 feet or less thereafter for scaffolds 3 feet wide or less, and every 26 feet or less thereafter for scaffolds greater than 3 feet wide.

3. The top guy, tie, or brace of completed scaffolds are to be placed no further than the 4:1 height from the top. Such guys, ties, and braces are to be installed at each end of the scaffold and at horizontal intervals not to exceed 30 feet (measured from one end [not both] toward the other).

4. Ties, guys, braces, or outriggers are to be used to prevent the tipping of supported scaffolds in all circumstances where an eccentric load, such as a cantilevered work platform, is applied or is transmitted to the scaffold.

Supported scaffold poles, legs, posts, frames, and uprights are to bear on base plates and mud sills or other adequate firm foundation (see Figure 96). Footings are to be level, sound, rigid, and capable of supporting the loaded scaffold without settling or displacement. Unstable objects are not to be used to support scaffolds or platform units, and are not to be used as working platforms. Supported scaffold poles, legs, posts, frames,

Figure 96. Scaffold legs and base plates supported on mud sills

and uprights are to be plumbed and braced to prevent swaying and displacement.

Front-end loaders and similar pieces of equipment are not to be used to support scaffold platforms unless they have been specifically designed by the manufacturer for such use. Fork-lifts are not to be used to support scaffold platforms unless the entire platform is attached to the fork and the fork-lift is not moved horizontally while the platform is occupied.

Suspension Scaffolds

All suspension scaffold support devices, such as outrigger beams, cornice hooks, parapet clamps, and similar devices, must rest on surfaces capable of supporting at least four times the load imposed on them by the scaffold operating at the rated load of the hoist, or at least 1.5 times the load imposed on them by the scaffold at the stall capacity of the hoist, whichever is greater. Suspension scaffold outrigger beams, when used, are to be made of structural metal or equivalent strength material, and are to be restrained to prevent movement.

The inboard ends of suspension scaffold outrigger beams are to be stabilized to the floor or roof deck by bolts or other direct connections, or they must have their inboard ends stabilized by counterweights; except masons' multipoint adjustable suspension scaffold outrigger beams are not to be stabilized by counterweights. Before the scaffold is used, direct connections are to be evaluated by a competent person who confirms, based on the evaluation, that the supporting surfaces are capable of supporting the loads to be imposed. In addition, masons' multipoint adjustable suspension scaffold connections are to be designed by an engineer experienced in such scaffold design.

Counterweights are to be made of nonflowable material. Sand, gravel, and similar materials that can be easily dislocated are not to be used as counterweights. Only those items specifically designed as counterweights are to be used to counterweight scaffold systems. Construction materials such as, but not limited to, masonry units and rolls of roofing felt, can be used as counterweights. Counterweights are to be secured by mechanical means to the outrigger beams to prevent accidental displacement. Counterweights are not to be removed from an outrigger beam until the scaffold is disassembled.

Outrigger beams, which are not stabilized by bolts or other direct connections to the floor or roof deck, are to be secured by tiebacks and the tiebacks are to be equivalent in strength to the suspension ropes. Outrigger beams are to be placed perpendicular to their bearing supports (usually the face of the building or structure). However, where the employer can demonstrate that it is not possible to place an outrigger beam perpendicular to the face of the building or structure because of obstructions that cannot be moved, the outrigger beam may be placed at some other angle, provided opposing angle tiebacks are used. Tiebacks are to be secured to a structurally sound anchorage on the building or structure. Sound anchorages include structural members, but do not include standpipes, vents, other piping systems, or electrical conduit.

Tiebacks are to be installed perpendicular to the face of the building or structure, or opposing angle tiebacks are to be installed. Single tiebacks which are installed at an angle are prohibited.

Suspension scaffold outrigger beams are to be

1.   Provided with stop bolts or shackles at both ends.

2.   Securely fastened together with the flanges turned out when channel iron beams are used in place of I-beams.

3.   Installed with all bearing supports perpendicular to the beam center line.

4.   Set and maintained with the web in a vertical position.

5.   When an outrigger beam is used, the shackle or clevis with which the rope is attached to the outrigger beam is to be placed directly over the center line of the stirrup.

Suspension scaffold support devices, such as cornice hooks, roof hooks, roof irons, parapet clamps, or similar devices, are to be

1.   Made of steel, wrought iron, or materials of equivalent strength.

2.   Supported by bearing blocks.

3.   Secured against movement by tiebacks installed at right angles to the face of the building or structure, or opposing angle tiebacks are to be installed and secured to a structurally sound point of anchorage on the building or structure. Sound points of anchorage include structural members, but do not include standpipes, vents, other piping systems, or electrical conduit.

4.   Tiebacks are to be equivalent in strength to the hoisting rope.

When winding drum hoists are used on a suspension scaffold, they are to contain not less than four wraps of the suspension rope at the lowest point of scaffold travel. When other types of hoists are used, the suspension rope is to be long enough to allow the scaffold to be lowered to the level below without the rope end passing through the hoist; or, the rope end must be configured or provided with a means to prevent the end from passing through the hoist. The use of repaired wire rope as suspension rope is prohibited. A wire suspension rope is not to be joined together except through the use of eye splice thimbles which are connected with shackles or coverplates and bolts. The load end of wire suspension ropes is to be equipped with proper size thimbles and secured by eyesplicing or equivalent means. Ropes are to be inspected for defects, by a competent person, prior to each workshift and after every occurrence which could affect a rope's integrity. Ropes are to be replaced if any of the following conditions exist:

1.   Any physical damage which impairs the function and strength of the rope.

2.   Kinks that might impair the tracking or wrapping of rope around the drum(s) or sheave(s).

3.   Six randomly distributed broken wires in one rope lay, or three broken wires in one strand in one rope lay.

4.   Abrasion, corrosion, scrubbing, flattening, or peening which causes loss of

more than one-third of the original diameter of the outside wires.

5.   Heat damage caused by a torch, or any damage caused by contact with electrical wires.

6.   Evidence that the secondary brake has been activated during an overspeed condition and has engaged the suspension rope.

Swagged attachments, or spliced eyes on wire suspension ropes, are not to be used unless they are made by the wire rope manufacturer or a qualified person. When wire rope clips are used on suspension scaffolds,

1.   There must be minimum of three wire rope clips installed, with the clips a minimum of six rope diameters apart.

2.   Clips are to be installed according to the manufacturer's recommendations.

3.   Clips are to be retightened to the manufacturer's recommendations after the initial loading.

4.   Clips are to be inspected and retightened to the manufacturer's recommendations at the start of each workshift thereafter.

5.   U-bolt clips are not to be used at the point of suspension for any scaffold hoist.

6.   When U-bolt clips are used, the U-bolt is to be placed over the dead end of the rope, and the saddle is to be placed over the live end of the rope.

Suspension scaffold power-operated hoists, and manual hoists, are to be tested by a qualified testing laboratory. Gasoline-powered equipment and hoists are not to be used on suspension scaffolds. Gears and brakes of power-operated hoists which are used on suspension scaffolds are to be enclosed. In addition to the normal operating brake, suspension scaffold power-operated hoists, and manually operated hoists, must have a braking device or locking pawl which engages automatically when a hoist makes either of the following uncontrolled movements: an instantaneous change in momentum, or an accelerated overspeed. Manually operated hoists require a positive crank force to descend.

Two-point and multipoint suspension scaffolds are to be tied or otherwise secured to prevent them from swaying; this is to be determined as necessary, based upon an evaluation by a competent person. Window cleaners' anchors are not to be used for this purpose. When devices are used for the sole purpose of providing emergency escape and rescue, they are not to be used as working platforms. This provision does not preclude the use of systems which are designed to function both as suspension scaffolds and emergency systems.

Accessing Scaffolds

When scaffold platforms are more than two feet above or below a point of access, portable ladders, hook-on ladders, attachable ladders, stair towers (scaffold stairways/towers), stairway-type ladders (such as ladder stands), ramps, walkways, integral prefabricated scaffold access, or direct access from another scaffold, structure, personnel hoist, or similar surface are to be used (see Figure 97). Crossbraces are not to be used as a means of access.

Portable, hook-on, and attachable ladders are to be positioned so as not to tip the scaffold. Hook-on and attachable ladders are to be positioned so that their bottom rung is not more than 24 inches above the scaffold supporting level. When hook-on and attachable ladders are used on a supported scaffold more than 35 feet high, they must have rest platforms at 35-foot maximum vertical intervals. Hook-on and attachable ladders are ones specifically designed for use with the type of scaffold used. They must have a minimum rung length of 11 1/2 inches and have uniformly spaced rungs with a maximum spacing between

Figure 97. Built-in scaffold access ladders

rungs of 16 3/4 inches.

Stairway-type ladders shall be positioned such that their bottom step is not more than 24 inches above the scaffold supporting level; shall be provided with rest platforms at 12 foot maximum vertical intervals; shall have a minimum step width of 16 inches (except that mobile scaffold stairway-type ladders must have a minimum step width of 11 1/2 inches); and shall have slip-resistant treads on all steps and landings.

Stairtowers (scaffold stairway/towers) are to be positioned such that their bottom step is not more than 24 inches above the scaffold supporting level.

Stairrails

A stairrail, which consists of a toprail and a midrail, is to be provided on each side of each scaffold stairway. The toprail of each stairrail system is also to be capable of serving as a handrail, unless a separate handrail is provided. Handrails, and toprails that serve as handrails, must provide an adequate handhold for employees grasping them to avoid falling. Stairrail systems and handrails are to be surfaced to prevent injury to employees from punctures or lacerations, and to prevent snagging of clothing. The ends of stairrail systems and handrails are to be constructed so that they do not constitute a projection hazard. Handrails, and toprails that are used as handrails, are to be at least 3 inches from other objects. Stairrails must not be less than 28 inches, nor more than 37 inches from the upper surface of the stairrail to the surface of the tread, and they must be in line with the face of the riser at the forward edge of the tread. A landing platform at least 18 inches wide by at least 18 inches long is to be provided at each level. Each scaffold stairway shall be at least 18 inches wide between stairrails. Guardrails shall be provided on the open sides and ends of each landing.

Stairways and Ramps

Stairways are to be installed between 40 degrees and 60 degrees from the horizontal. The riser height is to be uniform, within 1/4 inch, for each flight of stairs. Greater variations in riser height are allowed for the top and bottom steps of the entire system, not for each flight of stairs. Tread depth is to be uniform, within 1/4 inch, for each flight of stairs. Treads and landings are to have slip-resistant surfaces.

Ramps and walkways 6 feet or more above lower levels must have guardrail systems. No ramp or walkway is to be inclined more than a slope of one (1) vertical to three (3) horizontal (20 degrees above the horizontal.) If the slope of a ramp or a walkway is steeper than one (1) vertical in eight (8) horizontal, the ramp or walkway must have cleats not more than fourteen (14) inches apart, which are securely fastened to the planks to provide footing.

Integrated Scaffold Access

Integral prefabricated scaffold access frames are to be specifically designed and constructed for use as ladder rungs, and must have a rung length of at least 8 inches. They are not to be used as work platforms when rungs are less than 11 1/2 inches in length, unless each affected employee uses fall protection, or a positioning device. They must be uniformly spaced within each frame section and provide rest platforms at 35-foot maximum vertical intervals on all supported scaffolds more than 35 feet high and have a maximum spacing between rungs of 16 3/4 inches. Nonuniform rung spacing, caused by joining end frames together, is allowed provided the resulting spacing does not exceed 16 3/4 inches. Steps and rungs of ladders and stairway type access must line up vertically with each other between rest platforms. Direct access, to or from another surface, is to be used only when the scaffold is not more than 14 inches horizontally, and not more than 24 inches vertically from the other surface.

Access During Erecting or Dismantling

Where the provision of safe access is feasible and does not create a greater hazard, the employer is to provide a safe means of access for each employee erecting or dismantling a scaffold. The employer must have a competent person determine whether it is feasible to provide this access, or if it would pose a greater hazard. This determination is to be based on site conditions and the type of scaffold being erected or dismantled. Hook-on or attachable ladders are to be installed as soon as scaffold erection has progressed to a point that permits safe installation and use. When erecting or dismantling tubular welded frame scaffolds, (end) frames with horizontal members that are parallel, level, and are not more than 22 inches apart vertically, then these may be used as climbing devices for access, provided they are erected in a manner that creates a usable ladder which provides good hand hold and foot space. Cross braces on tubular welded frame scaffolds shall not be used as a means of access or egress.

Other Scaffold Rules

Scaffolds and scaffold components are not to be loaded in excess of their maximum intended loads or rated capacities, whichever is less. The use of shore or lean-to scaffolds is prohibited. Scaffolds and scaffold components are to be inspected for visible defects, by a competent person, before each work shift, and after any occurrence which could affect a scaffold's structural integrity. Any part of a scaffold that is damaged or weakened such that its strength is less than that required is to be immediately repaired or replaced, braced to meet those provisions, or removed from service until repaired.

Scaffolds are not to be moved horizontally while employees are on them, unless they have been designed by a registered professional engineer specifically for such movement.

Scaffolds are not to be erected, used, dismantled, altered, or moved so that they, or any conductive material handled on them, are close to exposed and energized power lines. Table16 provides clearance distances for scaffolds from power lines.

Table 16

Scaffold Clearance Distance From Power Lines

*Insulated Lines

Voltage	Minimum distance	Alternatives
Less than 300 volts. *300 volts to 50 kv. More than 50 kv.	3 feet (0.9 m) 10 feet (3.1 m) 10 feet (3.1 m) plus 0.4 inches (1.0 cm) for each 1 kv over 50 kv.	2 times the length of the line insulator, but never less than 10 feet (3.1 m).

*Uninsulated lines

Voltage	Minimum distance	Alternatives
Less than 50 kv. More than 50 kv.	10 feet (3.1 m). 10 feet (3.1 m) plus 0.4 inches (1.0 cm) or each 1 kv over 50 kv.	2 times the length of the line insulator, but never less than 10 feet (3.1 m).

Scaffolds and materials may be closer to power lines than specified above when such clearance is necessary for performance of work, but only after the utility company or electrical system operator has been notified of the need to work closer, and the utility company or electrical system operator has deenergized the lines, relocated the lines, or installed protective coverings to prevent accidental contact with the lines.

Scaffolds are to be erected, moved, dismantled, or altered only under the supervision and direction of a competent person qualified in scaffold erection, moving, dismantling, or alteration. Such activities are to be performed only by experienced and trained employees who are selected, by the competent person, for such work.

Employees are prohibited from working on scaffolds covered with snow, ice, or other slippery material, except as necessary for removal of such materials. Work on or from scaffolds is prohibited during storms or high winds unless a competent person has determined that it is safe for employees to be on the scaffold; those employees are to be protected by a personal fall arrest system or wind screens. Wind screens are not to be used unless the scaffold is secured against the anticipated wind forces imposed.

Where swinging loads are being hoisted onto, or near scaffolds so that the loads might contact the scaffold, tag lines or equivalent measures are to be used to control the loads.

Makeshift devices such as, but not limited to, boxes and barrels are not to be used on top of scaffold platforms to increase the working level height for employees. Ladders are not to be used on scaffolds to increase the working level height for employees, except on large area scaffolds when the ladder can be placed against a structure which is not a part of the scaffold. When this takes place, the scaffold is to be secured against the sideways thrust exerted by the ladder, and the platform units are to be secured to the scaffold to prevent their movement. The ladder legs are to be on the same platform, or other means are to be provided to stabilize the ladder against unequal platform deflection. Also, the ladder legs are to be secured to prevent them from slipping or being pushed off the platform.

To reduce the possibility of the welding current arcing through the suspension wire rope when performing welding from suspended scaffolds, the following precautions are to be taken, as applicable:

1. An insulated thimble is to be used to attach each suspension wire rope to its hanging support (such as cornice hook or outrigger). Excess suspension wire rope, and any additional independent lines from grounding, are to be insulated.

2. The suspension wire rope is to be covered with insulating material extending at least four feet above the hoist. If there is a tail line below the hoist, it is to be insulated to prevent contact with the platform. The portion of the tail line that hangs free below the scaffold shall be guided or retained, or both, so that it does not become grounded.

3. Each hoist is to be covered with insulated protective covers.

4. In addition to a work lead attachment which is required by the welding process, a grounding conductor is to be connected from the scaffold to the structure. The size of this conductor is to be at least the size of the welding process work lead, and this conductor is not to be in series with the welding process or the work piece.

5. If the scaffold grounding lead is disconnected at any time, the welding machine is to be shut off.

6. An active welding rod, or uninsulated welding lead, is not to be allowed to contact the scaffold or its suspension system.

Fall Protection

Each employee on a scaffold, which is more than 10 feet above a lower level, is to be protected from falling to that lower level. The fall protection requirements for employees installing suspension scaffold support systems on floors, roofs, and other elevated surfaces are set forth in Subpart M. Each employee on a boatswains' chair, catenary scaffold, float scaffold, needle beam scaffold, or ladder jack scaffold is to be protected by a personal fall arrest system. Workers on a single-point, or two-point adjustable suspension scaffold, are to be protected by both a personal fall arrest system and a guardrail system. Each employee on a crawling board (chicken ladder) is to be protected by a personal fall arrest system, a guardrail system (with minimum 200 pound toprail capacity), or by a three-fourth inch diameter grabline, or an equivalent handhold which is securely fastened beside each crawling board.

Each employee on a self-contained adjustable scaffold is to be protected by a guardrail system (with minimum 200 pound toprail capacity), when the platform is supported by the frame structure. Each employee is to be protected by both a personal fall arrest system and a guardrail system (with minimum 200 pound toprail capacity), when the platform is supported by ropes.

Each employee who is on a walkway that is located within a scaffold is to be protected by a guardrail system (with minimum 200 pound toprail capacity), and the guardrail system is to be installed within 9 1/2 inches of, and along at least one side of, the walkway.

Workers, who are performing overhand bricklaying operations from a supported scaffold, must be protected from falling from the open sides and ends of the scaffold (except at the side next to the wall being laid) by using a personal fall arrest system, or guardrail system (with minimum 200 pound toprail capacity).

For all other scaffolds, each employee is to use personal fall arrest systems, or guardrail systems for their protection. The employer must have a competent person determine the feasibility and safety of providing fall protection for employees erecting or dismantling supported scaffolds. Employers are required to provide fall protection for employees erecting or dismantling supported scaffolds, where the installation and use of such protection is feasible and does not create a greater hazard. Personal fall arrest systems used on scaffolds are to be attached by lanyard to a vertical lifeline, horizontal lifeline, or scaffold structural member.

When vertical lifelines are used, they are to be fastened to a fixed safe point of anchorage; they are to be independent of the scaffold; and they are to be protected from sharp edges and abrasion. Safe points of anchorage include structural members of buildings, but do not include standpipes, vents, other piping systems, electrical conduit, outrigger beams, or counterweights. Vertical lifelines are not to be used when overhead components, such as overhead protection or additional platform levels, are part of a single-point or two-point adjustable suspension scaffold. Vertical lifelines, independent support lines, and suspension ropes are not to be attached to each other; they are not to be attached to, or used at the same point of anchorage; nor are they to be attached to the same point on the scaffold or personal fall arrest system.

When horizontal lifelines are used, they are to be secured to two or more structural members of the scaffold, or they may be looped around both suspension and independent suspension lines (on scaffolds so equipped) above the hoist and the brake attached to the end of the scaffold. Horizontal lifelines are not to be attached only to the suspension ropes. When lanyards are connected to horizontal lifelines or structural members on a single-point or two-point adjustable suspension scaffold, the scaffold is to be equipped with additional independent support lines and automatic locking devices capable of stopping the fall of the scaffold, in the event one or both of the suspension ropes fail. The independent support lines are to be equal in number and strength to the suspension ropes.

Guardrail systems are to be installed along all open sides and ends of platforms. Guardrail systems are to be installed before the scaffold is released for use by employees other than erection/dismantling crews. The top edge height of toprails, or equivalent members on supported scaffolds which are manufactured or placed in service after January 1, 2000, shall be installed between 38 inches and 45 inches above the platform surface. The top edge height on supported scaffolds manufactured and placed in service before January 1, 2000, and on all suspended scaffolds where both a guardrail and a personal fall arrest system are required, are to be between 36 inches and 45 inches. When conditions warrant, the height of the top edge may exceed the 45-inch height.

When midrails, screens, mesh, intermediate vertical members, solid panels, or equivalent structural members are used, they are to be installed between the top edge of the guardrail system and the scaffold platform. When midrails are used, they are to be installed at a height approximately midway between the top edge of the guardrail system and the platform surface. When screens and mesh are used, they must extend from the top edge of the guardrail system to the scaffold platform, and along the entire opening between the supports. When intermediate members (such as balusters or additional rails) are used, they are not to be more than 19 inches apart. Each toprail, or equivalent member of a guardrail system, is to be capable of withstanding, without failure, a force applied in any downward or horizontal direction, at any point along its top edge, of at least 100 pounds for guardrail systems installed on single-point adjustable suspension scaffolds, or two-point adjustable suspension scaffolds. All other guardrail systems which are installed on all other scaffolds must be capable of withstanding at least 200 pounds. Midrails, screens, mesh, intermediate vertical members, solid panels, and equivalent structural members of a guardrail system are to be capable of withstanding, without failure, a force applied in any downward or horizontal direction, at any point along the midrail or other member, of at least 75 pounds for guardrail systems with a minimum 100 pound toprail capacity, and at least 150 pounds for guardrail systems with a minimum 200 pound toprail capacity. Suspension scaffold hoists and non-walk-through stirrups may be used as end guardrails, if the space between the hoist or stirrup and the side guardrail or structure does not allow passage of an employee to the end of the scaffold.

Guardrails are to be surfaced to prevent injury, to an employee, from punctures or lacerations, and to prevent snagging of clothing. Rails are not to overhang the terminal posts except when such overhang does not constitute a projection hazard to employees. Steel or plastic banding cannot be used as a toprail or midrail. Manila or plastic (or other synthetic) rope, being used for toprails or midrails, is to be inspected by a competent person, as frequently as necessary, to ensure that it continues to meet the strength requirements. Crossbracing is acceptable in place of a midrail, when the crossing point of two braces is between 20 inches and 30 inches above the work platform, or it may be used as a toprail when the crossing point of two braces is between 38 inches and 48 inches above the work platform. The end points at each upright are to be no more than 48 inches apart.

Falling Object Protection

In addition to wearing hardhats, each employee on a scaffold is to be provided with additional protection, from falling hand tools, debris, and other small objects, through the installation of toeboards, screens, or guardrail systems, or through the erection of debris nets, catch platforms, or canopy structures that contain or deflect the falling objects. When the falling objects are too large, heavy, or massive to be contained or deflected by any of the above-listed measures, the employer must place these potential falling objects away from the edge of the surface and must secure those materials, as necessary, to prevent their falling.

Where there is a danger of tools, materials, or equipment falling from a scaffold and striking employees below, the following provisions apply:

1. The area below the scaffold is to be barricaded, and employees are not to be permitted to enter the hazard area.

2. A toeboard is to be erected along the edge of platforms which are more than 10 feet above lower levels, and the distance is to be sufficient to protect employees below. On float (ship) scaffolds, an edging of 3/4 x 1 1/2 inch wood, or the equivalent, may be used in lieu of toeboards.

3. Where tools, materials, or equipment are piled to a height higher than the top edge of the toeboard, paneling or screening which extends from the toeboard or platform to the top of the guardrail is to be erected for a distance sufficient to protect employees below.

4. A guardrail system is to be installed with openings small enough to prevent passage of potential falling objects.

5. A canopy structure, debris net, or catch platform, strong enough to withstand the impact forces of the potential falling objects, is to be erected over the employees below.

Canopies, when used for falling object protection, are to be installed between the falling object hazard and the employees. For falling object protection, canopies are to be used on suspension scaffolds, and the scaffold is to be equipped with additional independent support lines, equal in number to the number of points supported, and equivalent in strength to the strength of the suspension ropes. Independent support lines and suspension ropes are not to be attached to the same points of anchorage.

Where used, toeboards are to be capable of withstanding, without failure, a force of at least 50 pounds applied in any downward or horizontal direction at any point along the toeboard, and at least three and one-half inches high from the top edge of the toeboard to the level of the walking/working surface. Toeboards are to be securely fastened in place at the outermost edge of the platform, and must have no more than 1/4 inch clearance above the walking/working surface. Toeboards are to be solid, or with openings not over one inch in the greatest dimension.

Additional Requirements Applicable to Specific Types of Scaffolds (1926.452)

Pole Scaffolds

When using pole scaffolds and platforms that are being moved to the next level, the existing platform is to be left undisturbed until the new bearers have been set in place and braced prior to receiving the new platforms. Crossbracing is to be installed between the inner and outer sets of poles on double pole scaffolds. Diagonal bracing, in both directions, is to be installed across the entire inside face of double-pole scaffolds, and it is to be used to support loads equivalent to a uniformly distributed load of 50 pounds or more per square foot. Also, diagonal bracing, in both directions, is to be installed across the entire outside face of all double- and single-pole scaffolds. All runners and bearers are to be installed on the edge. Bearers are to extend a minimum of 3 inches over the outside edges of runners, while runners must extend over a minimum of two poles, and are to be supported by bearing blocks which are securely attached to the poles. Braces, bearers, and runners are not to be spliced between poles. Where wooden poles are spliced, the ends are to be squared, and the upper section shall rest squarely on the lower section. Wood splice plates are to be provided on at least two adjacent sides, and they must extend at least 2 feet on either side of the splice, and overlap the abutted ends equally. They must have at least the same cross-sectional areas as the pole. Splice plates, of other materials of equivalent strength, may be used (see Figure 98).

Pole scaffolds over 60 feet in height are to be designed by a registered professional engineer, and are to be constructed and loaded in accordance with that design. Nonmandatory Appendix A of 29 CFR 1926.450 to Subpart L contains examples of criteria that will enable an employer to comply with the design and loading requirements for pole scaffolds under 60 feet in height.

Tubular and Coupler Scaffolds

On tublar and coupler scaffold when platforms are being moved to the next level, the existing platform is left undisturbed until the new bearers have been set in place and braced prior to receiving the new platforms. Transverse bracing, which forms an "X" across the width

Figure 98. Example of a pole scaffold

of the scaffold, is to be installed at the scaffold ends, at least at every third set of posts horizontally (measured from only one end), and at every fourth runner vertically. Bracing is to extend diagonally from the inner or outer posts or runners, upward to the next outer or inner posts or runners. Building ties are to be installed at the bearer levels between the transverse bracing. On straight run scaffolds, longitudinal bracing is to be installed across the inner and outer rows of posts, diagonally in both directions, and it must extend from the base of the end posts, upward to the top of the scaffold at approximately a 45 degree angle. On scaffolds, where the length is greater than their height, such bracing is to be repeated beginning at least at every fifth post. On scaffolds, where the length is less than their height, bracing is to be installed from the base of the end posts, upward to the opposite end posts, and then in alternating directions until reaching the top of the scaffold. Bracing is to be installed as close as possible to the intersection of the bearer and post or runner and post. Where conditions preclude the attachment of bracing to posts, bracing is to be attached to the runners as close to the post as possible.

Bearers are to be installed transversely between posts, and when coupled to the posts, they must have the inboard coupler bear directly on the runner coupler. When the bearers are coupled to the runners, the couplers are to be as close to the posts as possible. Bearers must extend beyond the posts and runners, and provide full contact with the coupler. Runners are to be installed along the length of the scaffold, and they are to be located on both the inside and outside posts at level heights (when tube and coupler guardrails and midrails are used on outside posts, they may be used in lieu of outside runners). Runners are to be interlocked on straight runs in order to form continuous lengths, and they are to be coupled to each post. The bottom runners and bearers are to be located as close to the base as possible.

Couplers are to be of a structural metal, such as drop-forged steel, malleable iron, or structural grade aluminum. The use of gray cast iron is prohibited. Tube and coupler scaffold, over 125 feet in height are to be designed by a registered professional engineer, and are to be constructed and loaded in accordance with such design. Nonmandatory Appendix A of 29 CFR 1926.450 to this subpart contains examples of criteria that will enable an employer to comply with design and loading requirements for tube and coupler scaffolds under 125 feet in height.

Fabricated Frame Scaffolds

On tubular welded frame scaffolds, when moving platforms to the next level, the existing platform is to be left undisturbed until the new end frames have been set in place and braced prior to receiving the new platforms. Frames and panels are to be braced by cross, horizontal, or diagonal braces, or combination thereof, which secure vertical members together laterally. Frames and panels are to be joined together vertically, by coupling or stacking pins, or by equivalent means. Where uplift can occur which would displace scaffold end frames or panels, the frames or panels are to be locked together vertically by pins or equivalent means.

The cross braces are to be of a length that will automatically square and align vertical members so that the erected scaffold is always plumb, level, and square. All brace connections are to be secured.

Brackets used to support cantilevered loads shall

1. Be seated with side-brackets parallel to the frames, and end-brackets at 90 degrees to the frames.

2. Not be bent or twisted from these positions.

3. Be used only to support personnel, unless the scaffold has been designed for other loads, by qualified engineer, and has been built to withstand the tipping forces caused by those other loads being placed on the bracket-supported section of the scaffold.

Scaffolds over 125 feet (38.0 m) in height above their base plates are to be designed by a registered professional engineer, and are to be constructed and loaded in accordance with such design.

Plasterers', Decorators', and Large Area Scaffolds

Plasterers', decorators', and large area scaffolds are to follow the construction guidelines and use which have been set forth for pole, tubular, coupler, and fabricated frame scaffolds.

Bricklayers' Square Scaffolds

Bricklayers' scaffolds are usually made of wood and are to be reinforced with gussets on both sides of each corner, and diagonal braces are to be installed on all sides of each square. Diagonal braces are to be installed between squares on the rear and front sides of the scaffold, and they must extend from the bottom of each square to the top of the next square. The scaffolds are not to exceed three tiers in height, and are to be so constructed and arranged that one square rests directly above the other. The upper tiers must stand on a continuous row of planks which are laid across the next lower tier, and they are to be nailed down, or otherwise secured, to prevent displacement.

Horse Scaffolds

Horse scaffolds are not to be constructed or arranged with more than two tiers, or be more than 10 feet in height, whichever is less. When horses are arranged in tiers, each horse is to be placed directly over the horse in the tier below. Thus, when the horses are arranged in tiers, the legs of each horse are to be nailed down, or otherwise secured to prevent displacement. If the horses are arranged in tiers, each tier is to be crossbraced.

Form Scaffolds and Carpenters' Bracket Scaffolds

All form or bracket scaffolds, except those for wooden bracket-form scaffolds, are to be attached to the supporting formwork or structure by means of one or more of the following: nails, a metal stud attachment device, welding, and by hooking over a secured structural supporting member with the form wales either bolted to the form, or secured by snap ties or tie bolts which extend through the form and are securely anchored. For carpenters' bracket scaffolds, only, they may be attached to the supporting formwork or structure by a bolt which extends through to the opposite side of the structure's wall.

Wooden bracket-form scaffolds are an integral part of the form panel. Folding type metal brackets, when extended for use, are either to be bolted or secured with a locking-type pin.

Roof Bracket Scaffolds

Scaffold brackets are to be constructed to fit the pitch of the roof and shall provide a level support for the platform. Brackets (including those provided with pointed metal projections) are to be anchored in place by nails, unless it is impractical to use nails. When nails are not used, brackets are to be secured in place with first-grade manila rope of at least three-fourth inch diameter, or equivalent (see Figure 99).

Figure 99. Example of roof bracket scaffolds

Outrigger Scaffolds

The inboard end of outrigger beams of an outrigger scaffold, measured from the fulcrum point to the extreme point of anchorage, is not to be less than one and one-half times the outboard end in length. Outrigger beams are to be fabricated in the shape of an I-beam or channel, and are to be placed so that the web section is vertical. The fulcrum point of outrigger beams must rest on secure bearings at least 6 inches in each horizontal dimension. Outrigger beams are to be secured in place against movement, and are to be securely braced at the fulcrum point against tipping. The inboard ends of outrigger beams are to be securely anchored, either by means of braced struts bearing against sills which are in contact with the overhead beams or ceiling, or by means of tension members which are secured to the floor joists underfoot, or by both. The entire supporting structure must be securely braced to prevent any horizontal movement. To prevent their displacement, platform units are to be nailed, bolted, or otherwise secured to outriggers. The scaffolds and scaffold components are to be designed by a registered professional engineer, and are to be constructed and loaded in accordance with such design.

Pump Jack Scaffolds

Pump jack brackets, braces, and accessories are to be fabricated from metal plates and angles. Each pump jack bracket must have two positive gripping mechanisms to prevent any failure or slippage. Poles are to be secured to the structure by rigid triangular bracing or equivalent at the bottom, top, and other points, as necessary. When the pump jack has to pass bracing already installed, an additional brace is to be installed approximately 4 feet above the brace to be passed, and it is to be left in place until the pump jack has been moved and the original brace reinstalled. When guardrails are used for fall protection, a workbench may be used as the toprail (see Figure 100). Work benches are not to be used as scaffold platforms.

Figure 100. Pump jack scaffold

When poles are made of wood, the pole lumber is to be straight-grained, free of shakes, large loose or dead knots, and other defects which might impair strength. If the wood poles are constructed of two continuous lengths, they are to be joined together with the seam parallel to the bracket. When two by fours are spliced to make a pole, mending plates are to be installed at all splices in order to develop the full strength of the member.

Ladder Jack Scaffolds

On ladder jack scaffold platforms, do not exceed a height of 20 feet. All ladders used to support ladder jack scaffolds must meet the requirements of Subpart X of this part – Stairways and Ladders, except job-made ladders are not to be used to support ladder jack scaffolds. The ladder jack is to be so designed and constructed that it will bear on the side rails and ladder rungs, or on the ladder rungs alone. If bearing on rungs only, the bearing area must include a length of at least 10 inches on each rung. Ladders used to support ladder jacks are to be placed, fastened, or equipped with devices to prevent slipping. Scaffold platforms are never to be bridged one to another.

Window Jack Scaffolds

Window jack scaffolds are to be securely attached to the window opening. These scaffolds are to be used only for the purpose of working at the window opening through which the jack is placed. Window jacks are not to be used to support planks placed between one window jack and another, or for other elements of scaffolding.

Crawling Boards (Chicken Ladders)

Crawling boards must extend from the roof peak to the eaves, when used in connection with roof construction, repair, or maintenance. Crawling boards are to be secured to the roof by ridge hooks, or by a means that meets equivalent criteria (e.g., strength and durability).

Step, Platform, and Trestle Ladder Scaffolds

Step, platform, and trestle ladder scaffolds are not to be placed any higher than the second highest rung or step of the ladder supporting the platform. All ladders used in conjunction with step, platform, and trestle ladder scaffolds must meet the pertinent requirements of Subpart X of this part – Stairways and Ladders, except that job-made ladders are never to be used to support such scaffolds. Ladders used to support step, platform, and trestle ladder scaffolds are to be placed, fastened, or equipped with devices to prevent slipping, and these scaffolds shall not be bridged one to another.

Single-Point Adjustable Suspension Scaffolds

When two single-point adjustable suspension scaffolds are combined to form a two-point adjustable suspension scaffold, the resulting two-point scaffold must comply with the requirements for two-point adjustable suspension scaffold.

On a single-point adjustable suspension scaffold, the supporting rope between the scaffold and the suspension device is to be kept vertical unless all of the following conditions are met:

1. The rigging has been designed by a qualified person.

2. The scaffold is accessible to rescuers.

3. The supporting rope is protected to ensure that it will not chafe at any point where a change in direction occurs.

4. The scaffold is positioned so that swinging cannot bring the scaffold into contact with another surface.

Boatswains' Chair

Boatswains' chair tackle must consist of the correct size ball bearings or bushed blocks which contain safety hooks, and must be properly "eye-spliced" with a minimum five-eighth (5/8) inch diameter first-grade manila rope, or other rope which will satisfy the criteria (e.g., strength and durability) of manila rope. Boatswains' chair seat slings are to be reeved through four corner holes in the seat; cross each other on the underside of the seat; and are to be rigged so as to prevent slippage that could cause an out-of-level condition. Boatswains' chair seat slings are to be a minimum of five-eight (5/8) inch diameter fiber, synthetic, or other rope which will satisfy the criteria (e.g., strength, slip resistance, durability, etc.) of first grade manila rope. When a heat-producing process such as gas or arc welding is being conducted, boatswains' chair seat slings are to be a minimum of three-eight (3/8) inch wire rope. Non-cross-laminated wood boatswains' chairs are to be reinforced on their underside by cleats securely fastened to prevent the board from splitting.

Two-Point Adjustable Suspension Scaffolds (Swing Stages)

To prevent unstable conditions, two-point adjustable suspension scaffolds' (swing stages) platforms are not to be more than 36 inches wide, unless designed by a qualified person. The platforms are to be securely fastened to hangers (stirrups) by U-bolts, or by other satisfactory means. The blocks for fiber or synthetic ropes must consist of at least one double and one single block. The sheaves of all blocks shall fit the size of the rope to be used.

Platforms are to be of the ladder-type, plank-type, beam-type, or light-metal type. Light metal-type platforms having a rated capacity of 750 pounds or less, and platforms 40 feet or less in length which are tested and listed by a nationally recognized testing laboratory, are to be used.

Two-point scaffolds are not to be bridged or otherwise connected one to another during raising and lowering operations, unless the bridge connections are articulated (attached), and the hoists are properly sized. Passage may be made from one platform to another only when the platforms are at the same height, are abutting, and walk-through stirrups, which are specifically designed, are used for this purpose.

Multipoint Adjustable Suspension Scaffolds, Stonesetters' Multipoint Adjustable Suspension Scaffolds, and Masons' Multipoint Adjustable Suspension Scaffolds

Multipoint adjustable suspension scaffolds, stonesetters' multipoint adjustable suspension scaffolds, and masons' multipoint adjustable suspension scaffolds which use two or more scaffolds, are not to be bridged one to another unless they are designed to be bridged. The bridge connections are to be articulated, and the hoists are to be properly sized. If bridges are not used, passage may be made from one platform to another only when the platforms are at the same height and are abutting. Scaffolds are to be suspended from metal outriggers, brackets, wire rope slings, hooks, or by means that meet equivalent criteria (e.g., strength, durability).

Catenary Scaffolds

No more than one platform is to be placed between consecutive vertical pickups, and no more than two platforms are to be used on a catenary scaffold. Platforms, supported by wire ropes, must have hook-shaped stops on each end of the platforms to prevent them from slipping off the wire ropes. These hooks are to be placed so that they will prevent the platform from falling if one of the horizontal wire ropes breaks. Wire ropes must not be tightened to the extent that the application of a scaffold load will overstress them. Wire ropes are to be continuous and without splices between anchors.

Float (Ship) Scaffolds

On a float scaffold, the platform is to be supported by a minimum of two bearers, each of which must project a minimum of 6 inches beyond the platform on both sides. Each bearer is to be securely fastened to the platform. Any rope connections are to be such that the platform cannot shift or slip. When only two ropes are used with each float, they are to be arranged so as to provide four ends which are securely fastened to overhead supports. Each supporting rope is to be hitched around one end of the bearer and pass under the platform to the other end of the bearer, where it is hitched again, leaving sufficient rope at each end for the supporting ties.

Interior Hung Scaffolds

Interior hung scaffolds are to be suspended only from the roof structure or other structural member, such as ceiling beams. Overhead supporting members (roof structure, ceiling beams, or other structural members) are to be inspected and checked for strength before the scaffold is erected. Suspension ropes and cables are to be connected to the overhead supporting members by shackles, clips, thimbles, or other means that meet equivalent criteria (e.g., strength, durability).

Needle Beam Scaffolds

These scaffold's support beams are to be installed on edge. Ropes or hangers are to be used for supports, except that one end of a needle beam scaffold may be supported by a permanent structural member. The ropes are to be securely attached to the needle beams. The support connection is to be arranged so as to prevent the needle beam from rolling or becoming displaced. Platform units are to be securely attached to the needle beams by bolts or equivalent means. Cleats and overhang are not considered to be adequate means of attachment.

Multi-Level Suspended Scaffolds

Multi-level suspended scaffolds are to be equipped with additional independent support lines, equal in number to the number of points supported, and of equivalent strength to the suspension ropes. They must be rigged to support the scaffold in the event the suspension rope(s) fail. Independent support lines and suspension ropes are not to be attached to the same points of anchorage. Supports for platforms are to be attached directly to the support stirrup and not to any other platform.

Mobile Scaffolds

Mobile scaffolds must be braced by cross, horizontal, or diagonal braces, or a combination thereof, to prevent racking or collapse of the scaffold, and to secure vertical members together laterally so as to automatically square and align the vertical members (See Figure 101). Scaffolds are to be plumb, level, and squared. All brace connections are to be secured.

Scaffolds constructed of tube and coupler or fabricated frame components must comply with the previous requirements for these types of scaffolds. When scaffolds are equipped with casters and wheels, they are to be the locking type with positive wheel and/or wheel and swivel locks, or equivalent means, to prevent movement of the scaffold while the scaffold is used in a stationary manner.

The use of manual force to move the scaffold is to be applied as close to the base as practicable, but not more than 5 feet above the supporting surface. Power systems used to propel mobile scaffolds are to be designed for such use. Forklifts, trucks, similar motor vehicles, or add-on motors are not to be used to propel scaffolds unless the scaffold is designed for such propulsion systems.

Scaffolds are to be stabilized to prevent tipping during movement. Workers are not allowed to ride on scaffolds unless the following conditions exist:

1. The surface on which the scaffold is being moved is within 3 degrees of level, and free of pits, holes, and obstructions.

2. The height-to-base width ratio of the scaffold during movement is two to one or less, unless the scaffold is designed and constructed to meet or exceed nationally recognized stability test requirements.

Figure 101. Example of a mobile scaffold

3.   Outrigger frames, when used, are to be installed on both sides of the scaffold.

4.   When power systems are used, the propelling force is to be applied directly to the wheels, and it must not produce a speed in excess of one foot per second.

5.   No employee is to be on any part of the scaffold which extends outward beyond the wheels, casters, or other supports.

Platforms must not extend outward beyond the base supports of the scaffold, unless outrigger frames or equivalent devices are used to ensure stability. Where leveling of the scaffold is necessary, screw jacks or equivalent means are to be used. Caster stems and wheel stems are to be pinned or otherwise secured in scaffold legs or adjustment screws. Before a scaffold is moved, each employee on the scaffold is to be made aware of the move.

<u>Repair Bracket Scaffolds</u>

Brackets are to be secured in place by at least one wire rope at least 1/2 inch in diameter. Each bracket is to be attached to the securing wire rope (or ropes) by a positive locking device capable of preventing the unintentional detachment of the bracket from the rope, or by equivalent means. At the contact point between the supporting structure and the bottom of the bracket, a shoe (heel block or foot) must be provided which is capable of preventing the lateral movement of the bracket. Platforms are to be secured to the brackets in a manner that will prevent the separation of the platforms from the brackets and the movement of the platforms or the brackets on a completed scaffold.

When a wire rope is placed around the structure in order to provide a safe anchorage for personal fall arrest systems used by employees erecting or dismantling scaffolds, the wire rope is to be at least 5/16 inch in diameter. Each wire rope used for securing brackets in place, or used as an anchorage for personal fall arrest systems, is to be protected from damage due to contact with edges, corners, protrusions, or other discontinuities of the supporting structure or scaffold components. Tensioning of each wire rope used for securing brackets in place, or used as an anchorage for personal fall arrest systems, is to be by means of a turnbuckle at least 1 inch in diameter, or by equivalent means. Each turnbuckle is to be connected to the other end of its rope by use of an eyesplice thimble of a size appropriate to the turnbuckle to which it is attached. U-bolt wire rope clips are not to be used on any wire rope used to secure brackets, or used to serve as an anchor for personal fall arrest systems. The employer must ensure that materials are not dropped to the outside of the supporting structure. Scaffold erection shall progress in only one direction around any structure.

Stilts

An employee may wear stilts on a scaffold only if it is a large area scaffold. When an employee is using stilts on a large area scaffold where a guardrail system is used to provide fall protection, the guardrail system is to be increased in height by an amount equal to the height of the stilts being used by the employee. Surfaces on which stilts are used are to be flat and free of pits, holes, and obstructions such as debris, as well as other tripping and falling hazards. Stilts are to be properly maintained and any alteration of the original equipment is to be approved by the manufacturer.

Training Requirements (1926.454)

The employer must assure that each employee who performs work while on a scaffold is trained by a person qualified in the subject matter to recognize the hazards associated with the type of scaffold being used, and to understand the procedures to control or minimize those hazards. The training shall include the following areas, as applicable:

1. The nature of any electrical hazards, fall hazards, and falling object hazards in the work area.

2. The correct procedures for dealing with electrical hazards, and the procedures for erecting, maintaining, and disassembling the fall protection and falling object protection systems being used.

3. The proper use of the scaffold, and the proper handling of materials on the scaffold.

4. The maximum intended load, and the load-carrying capacities of the scaffolds used.

5. Any other pertinent requirements or information.

The employer must have each employee, who is involved in erecting, disassembling, moving, operating, repairing, maintaining, or inspecting a scaffold, trained, by a competent person, to recognize any hazards associated with the work in question. The training shall include the following topics, as applicable:

1. The nature of scaffold hazards.

2. The correct procedures for erecting, disassembling, moving, operating, repairing, inspecting, and maintaining the type of scaffold in question.

3. The design criteria, maximum intended load-carrying capacity, and intended use of the scaffold.

4. Any other pertinent requirements or information.

When the employer has reason to believe that an employee lacks the skill or understanding needed to work safely with the erection, use, or dismantling of scaffolds, the employer must retrain the employee so that the requisite proficiency is regained. Retraining is to be required in at least the following situations: where changes at the worksite present a hazard;

when an employee has not been previously trained in the types of scaffolds, fall protection, falling object protection, or other equipment that presents a hazard; and when an employer notices inadequacies in an employee who is working on scaffolds and there is an indication that the employee has not retained the requisite proficiency.

SIGNS, SIGNALS, AND BARRICADES

Construction activities at the jobsite may present several potential hazards to workers. The use of signs, signals, and barricades is essential in order to make employees aware that an immediate or potential hazard exists (see Figure 102). Therefore, the following safe work procedures for signs, signals, and barricades must be implemented and enforced on each construction project.

Figure 102. Warning sign for site requirements

Accident Prevention Signs/Tags

1. **Danger Signs** must be used wherever an immediate hazard (i.e., electrical conductor) exists. The danger signs must have red as the predominant color, in the upper panel, and a white lower panel, for additional sign wording (see Figure 103).

2. **Caution Signs** must be used to warn against potential hazards, or to caution against unsafe practices. The caution signs must have yellow as the predominant color, with a black upper panel (yellow lettering of "caution" on the upper panel), and a yellow lower panel, for additional sign wording.

3. **Exit Signs,** when required, shall be in legible red letters, not less than 6 inches high, on a white field.

4. **Safety Instruction Signs**, when used, must be white with a green upper panel, and white lettering to convey the principal message. Any additional wording must be in black lettering on the white background.

Figure 103. An example of an electrical danger sign

5. **Directional Signals** must be white with a black panel, and a white directional symbol. Any additional wording must be in black lettering on a white background.

6. **Traffic Signs** must be posted, at points of hazards, in all construction areas. All traffic control signs or devices must conform to ANSI D6.1-1971, *Manual on Uniform Traffic Control Devices for Streets and Highways.*

7. **Accident Prevention Tags** will be used as a temporary means of warning employees of an existing hazard, such as defective tools, equipment, etc.

8. **Out of Order Tags** will be used to designate equipment which requires repair or maintenance. Equipment with such a tag may not be used until the tag is removed.

9. Additional rules, not specifically prescribed in this section, are contained in ANSI Z35.1 1968, Specifications for Accident Prevention Signs and Z35.2-1968, Specifications for Accident Prevention Tags.

SITE CLEARANCE (1926.604)

When workers are engaged in clearing a site, they must be protected from hazards of irritant and toxic plants, and they must be suitably instructed in the first aid treatment available. All equipment used in site clearing operations is to be equipped with rollover protection. In addition, rider-operated equipment is to be equipped with an overhead and rear canopy guard which meets the following requirements:

1. The overhead covering on this canopy structure is not to be less than an 1/8-inch steel plate, or a 1/4-inch woven wire mesh, with openings no greater than 1 inch, or equivalent.

2. The opening in the rear of the canopy structure is to be covered with not less than 1/4-inch woven wire mesh, with openings no greater than 1 inch.

SLIPS, TRIPS, AND FALLS

Slips, trips, and falls result in approximately 20% of workers' compensation claims, 15% of all disabling injuries, 69% of fall-to-same-level injuries, and 15% of occupational deaths due to falls.

The causes for slips, trips, and falls have been linked to an unsafe mindset, unsafe conditions, and unsafe behaviors. Many of these types of accidents could be prevented by becoming aware of state of mind, identifying unsafe conditions or behaviors, selecting the right tools for the job, and using the correct body mechanics. Some examples of using good body mechanics are

1.   Don't tilt head.

2.   Use all your fingers to grip.

3.   Shorten your stride and point feet outward slightly.

4.   Walk with knees slightly bent which will help to avoid falling forward.

5.   Balance any load you carry.

6.   Avoid reaching too far.

Maintain a good center of balance by using eyes, ears, and muscles; correct vision problems; stay in shape; maintain a normal weight; and do not use alcohol or drugs.

Also, some time must be spent looking for potential hazards, such as loose or bent boards, broken floor tiles, unsecured mats or carpets, floor surfaces that change elevation, broken concrete, uncovered manholes, uncovered drains, unsafe ladders or stairs, slippery surfaces, obstructions in walkways, poor lighting, the use of improper shoes, and running or moving too fast. Some hazards and remedies can be found in Table 17.

Table 17

Making the Work Area Safe

HAZARD	REMEDY
Wet/uneven surfaces	Clean up/report spills
Improper lighting	Keep areas well lit
Winter weather	Wear the right shoes
Wrong tools	Use job-appropriate tools
Poor housekeeping	Clean clutter
Big loads	Carry smaller loads
Not enough time	Slow down

Some shoes are better than others for certain sets of circumstances. Soles which are better for wet surfaces are those that are synthetic rubber soles (neoprene); they should not be worn when oil exists on the surface. Crepe soles (crinkle rubber) are good on rough concrete, but not on tile, wood, or smooth concrete; for these surfaces, aluminum oxide bonded soles are appropriate. For dry surfaces, the use of neoprene soles, crepe soles, or soft rubber soles work well. Hard rubber soles are good for oily surfaces, but are not good for tile, wood surfaces, or wet or dry concrete.

Stairways are often culprits in causing slips, trips, and falls. Some of the stairways' most common conditions and hazards are: stairways without handrails; tools, equipment, litter, and spills left on stairways; stairways covered with ice or snow; and stairways not fully constructed.

Most workers often trip or stumble over unexpected objects in their way; thus, the importance of housekeeping is a key strategy in the prevention of slips, trips, and falls. (see Table 18).

Table 18

Housekeeping

HOUSEKEEPING DO'S

– Keep everything at work in its proper place.

– Put things away after use.

– Have adequate lighting, or use a flashlight.

– Walk and change directions slowly, especially when carrying anything.

– Make sure the teeth or head on a wrench is in good shape and won't slip when you pull on it.

HOUSEKEEPING DON'TS

– Don't leave machines, tools, or other materials on the floor.

– Don't block walkways or aisles with machines or equipment.

– Don't use a "cheater" on a wrench; get a larger wrench with a larger head or handle, if you need it.

– Don't leave cords, power cables, or air hoses in walkways.

– Don't place anything on stairs.

– Don't leave drawers open.

– Don't carry or push loads that block your vision.

When falls to the same level happen, it usually occurs because someone is running or walking too fast; dangerous surfaces exist, such as icy or wet surfaces; poor visibility exists due to dust, glare, or smoke; or because a worker is carrying a load that blocks the worker's vision.

Clothing can be, at times, a contributing factor to an accident. A fall can occur when a worker stumbles over a loose pant cuff, or when the shoes the worker is wearing aren't appropriate for the job or activity. A worker who is not watching where he or she is going has a greater potential of having an accident, as well as does a worker who does not keep all four legs on the floor when sitting on a chair.

In order to prevent slips, the following guidelines can be followed:

1. Clean up spills, drips, and leaks immediately.

2. Sand icy spots immediately — and tread carefully.

3. Use slip-resistant floor waxes and polishes in offices.

4.   Use steel drains, grates, and splash guards.

5.   Use non-slip paint, mats, treads, and abrasive surfaces to roughen plates, grills, and concrete walking surfaces.

6.   Use rough or grained steel surfaces in areas where there are often spills.

7.   Put up signs or barriers to keep workers away from temporary slip hazards.

8.   Wear shoes with anti-skid soles and materials that resist oils and acids; ask your supervisor about the right shoes for the job.

9.   Avoid turning sharply when walking on a slippery surface.

10. Keep hands at your sides, not in your pockets.

11. Walk slowly and slide feet when on wet, slippery, or uneven surfaces.

12. Don't count on the other workers to report the hazard.

13. Most important, be careful and take responsibility for your own actions.

Workers often slip, trip, or fall because they are busy performing other tasks and do not pay attention to the risks and the existing hazards. Although there are no foolproof ways to prevent slips, trips, and falls, there is much that can be done, by everyone on a construction site, to mitigate the hazards and foster a mindset which can prevent these types of accidents.

STAIRWAYS (1926.1052)

A stairway or ladder is to be provided, at all personnel points of access, where there is a break in elevation of 19 inches or more, and where there is no ramp, runway, sloped embankment, or personnel hoist provided. Workers are not to use any spiral stairways that will not be a permanent part of the structure on which construction work is being performed.

Stairways that will not be a permanent part of the structure on which construction work is being performed must have landings of not less than 30 inches in the direction of travel, and must extend at least 22 inches in width at every 12 feet or less of vertical rise. Stairs are to be installed between 30 degrees and 50 degrees from the horizontal. Riser height and tread depth must be uniform within each flight of stairs, including any foundation structure used as one or more treads of the stairs. Variations in riser height or tread depth shall not be over 1/4-inch in any stairway system.

Where doors or gates open directly on a stairway, a platform is to be provided, and the swing of the door is not to reduce the effective width of the platform to less than 20 inches.

Metal pan landings and metal pan treads, when used, are to be secured in place before filling with concrete or other material (see Figure 104). Except during stairway construction, foot traffic is prohibited on stairways with pan stairs, where the treads and/or landings are to be filled in with concrete or other material at a later date, unless the stairs are temporarily fitted with wood or other solid material at least to the top edge of each pan. Such temporary treads and landings are to be replaced when worn below the level of the top edge of the pan. Metal pan landings and treads should not be left empty since this results in a severe tripping hazard (see Figure 105).

All parts of stairways are to be free of hazardous projections, such as protruding nails. Slippery conditions on stairways are to be eliminated before the stairways are used to reach other levels. Except during stairway construction, foot traffic is prohibited on skeleton metal stairs where permanent treads and/or landings are to be installed at a later date, unless the stairs are fitted with secured temporary treads and the landings are long enough to cover the entire

Figure 104. Stair pans filled with wood

tread and/or landing area. Treads for temporary service can be made of wood or other solid material, and are to be installed the full width and depth of the stair.

Stairways having four or more risers, or rising more than 30 inches, whichever is less, are to be equipped with at least one handrail and one stairrail system, along each unprotected side or edge. Handrails and the top rails of stairrail systems are to be capable of withstanding, without failure, a force of at least 200 pounds applied within 2 inches of the top edge, in any downward or outward direction, at any point along the top edge.

Figure 105. Unfilled stair pans create a real slip/trip/fall hazard

The height of handrails are not to be more than 37 inches, nor less than 30 inches from the upper surface of the handrail to the surface of the tread, and are to be in line with the face of the riser at the forward edge of the tread. Handrails must provide an adequate handhold for employees to grasp them to avoid falling. Handrails that will not be a permanent part of the structure being built must have a minimum clearance of 3 inches between the handrail and walls, stairrail systems, and other objects. Winding and spiral stairways are to be equipped with a handrail, offset sufficiently, to prevent walking on those portions of the stairways where the tread width is less than 6 inches.

The height of stairrails are not to be less than 36 inches from the upper surface of the stairrail system to the surface of the tread, and they must be in line with the face of the riser at the forward edge of the tread. Stairrails installed before March 15, 1991 are not to be less than 30 inches, nor more than 34 inches from the upper surface of the stairrail system to the surface of the tread, and they must be in line with the face of the riser at the forward edge of the tread. Midrails, screens, mesh, intermediate vertical members, or equivalent intermediate structural members, are to be provided between the top rail of the stairrail system and the stairway steps. Midrails, when used, are located at a height midway between the top edge of the stairrail system and the stairway steps.

Screens or mesh, when used, must extend from the top rail to the stairway step, and along the entire opening between top rail supports. When intermediate vertical members, such as balusters, are used between posts, they are not to be more than 19 inches apart. Other structural members, when used, are to be installed such that there are no openings in the stairrail system that are more than 19 inches wide.

The ends of stairrail systems and handrails are to be constructed so as not to constitute a projection hazard. Stairrail systems and handrails are to be surfaced to prevent injury to employees from punctures or lacerations, and to prevent snagging of clothing.

When the top edge of a stairrail system also serves as a handrail, the height of the top edge is not to be more than 37 inches, nor less than 36 inches from the upper surface of the stairrail system to the surface of the tread, and it must be in line with the face of the riser at the forward edge of the tread.

Unprotected sides and edges of stairway landings must be provided with guardrail systems.

When using stairways, hold the handrails when ascending or descending, maintain a good path for vision, if carrying objects while on stairs, and watch your footing in poorly lighted stairwells, or where there are spills, ice, or snow on the stairs (see Table 19 and Table 20).

Table 19

Unsafe Acts on Stairways

- Climbing or descending stairs without holding onto the handrail.
- Carrying a load, especially one that blocks visibility.
- Not cleaning known slippery surfaces.
- Lack of concentration.
- Not keeping stairs free of clutter.
- Forgetting or ignoring safe work practices.
- Slow physical reaction, dizziness, or vision problems.

Table 20

Stairs Checklist

___	1. Do all stairs have handrails?
___	2. Are stairs constructed of sound materials — free of loose steps, splits, cracks, warping, rust, corrosion, or other defects?
___	3. Are spills cleaned up immediately?
___	4. Are stairs free of ice or snow?
___	5. Is the jobsite equipped with lifting devices that keep load carrying on stairs to a minimum?
___	6. Do workers go up and down stairs without using handrails?
___	7. Do workers carry heavy loads while going up and down stairs?
___	8. Do workers carry loads which block their vision while going up and down stairs?
___	9. Do workers run up or down stairs?
___	10. Do workers go up or down stairs cluttered with tools, equipment, litter, or personal belongings?
___	11. Do workers go up or down stairs covered with ice or snow?

STEEL ERECTION (1926.750)

Although erecting steel is one of the unique skills that steel erectors or ironworkers must have, it is still one of the most dangerous activities for construction workers (see Figure 106). OSHA has been actively working at coming up with a revised regulation regarding steel erection. See Table 21 for a review of the highlights of the proposed changes. The performing of steel erections has resulted in an average of 28 deaths and 1,800 lost-workday injuries each year. The existing hazards include working under loads, hoisting, landing and placing decking, column stability, double connections, landing and placing steel joists, and falling to lower levels.

Figure 106. A steel worker's precarious perch

Table 21

Highlights of SENRAC Consensus Proposal for Steel Erection

Site Layout and Construction Sequence
- Requires certification of proper curing of concrete in footings, piers, etc., for steel columns.
- Requires controlling contractor to provide erector with a safe site layout, including preplanning routes for hoisting loads.

Site-Specific Erection Plan
- Requires preplanning of key erection elements, including coordination with controlling contractor before erection begins, in certain circumstances.

Hoisting and Rigging
- Provides additional crane safety for steel erection by including elements of American National Standards Institute (ANSI) B30.5 - 1994 in pre-shift inspection.
- Minimizes employee exposure to overhead loads through preplanning and work practice requirements.
- Prescribes proper procedures for multiple lifts ("Christmas-treeing") that reduces overhead exposure, operator fatigue, and eliminates improper work practices.

Structural Steel Assembly
- Provides safer walking/working surfaces by eliminating tripping hazards, such as shear connectors, and minimizes slips through new slip resistance requirements.
- Provides specific work practices regarding landing deck bundles safely, and promotes prompt protection from fall hazards in interior openings.

Anchor Bolts
- Eliminates a major cause of collapse by requiring four anchor bolts per column, along with other column stability requirements.
- Requires procedures that assure the adequacy of anchor bolts which have been modified in the field.

Beams and Columns
- Eliminates the extremely dangerous collapse hazards associated with making double connections at columns by prescribing a safe procedure.

Open Web Steel Joists
- Minimizes the hazards associated with collapse of lightweight steel joists by addressing the need for erection bridging, and a method of attachment.
- Requires bridging terminus anchors as a performance standard, with illustrations and drawings provided in a non-mandatory appendix.
- Minimizes collapse hazards, while placing loads on steel joists, with new requirements that reflect proper industry procedures.

Pre-Engineered Metal Buildings
- Minimizes collapse hazards in the erection of these specialized structures, which account for the majority of steel erection in the U.S.

Falling Object Protection
- Provides protection against all hazards of falling objects in steel erection.

Table 21

Highlights of SENRAC Consensus Proposal for Steel Erection (Continued)

Fall Protection
- Requires fall protection at heights greater than 15 feet, with exceptions for connectors and deckers working in a controlled decking zone. A controlled decking zone is an area where work, such as the initial installation and placement of the metal deck, may be performed without the use of guardrail systems, personal fall arrest systems, or safety net systems, and where access to the zone is controlled.
- Connectors and deckers in the controlled decking zone must be protected at heights greater than two stories, or 30 feet. Connectors, working between 15 and 30 feet, must wear fall arrest or restraint equipment, and those workers be able to be tied off or be provided with another means of fall protection.
- Requires additional provisions for controlled decking zones, because of the number of decking fatalities which have been reported.

Training
- Requires a qualified person to train exposed workers in fall protection.
- Requires a qualified person to train exposed workers engaged in special activities, such as "Christmas treeing," connecting, and decking.

Floor Requirements (1926.750)

During steel erection the installation of permanent floors, as the erection of structural members progresses, and is not to be more than eight stories between the erection floor and the uppermost permanent floor, except where the structural integrity is maintained as a result of the design (see Figure 107). There is not to be more than four floors, or 48 feet of unfinished bolting or welding above the foundation, or uppermost permanently secured floor.

Figure 107. Flooring installed during steel erection

All derrick or erection floor is to be solidly planked or decked over its entire surface, except for access openings. Planking or decking, of equivalent strength, must be of the proper thickness to carry the working load. Planking is not to be less than 2 inches thick full size undressed, and is to be laid tight, and secured to prevent movement.

If a building or structure is not adaptable to temporary floors, and where scaffolds are not used, safety nets are to be installed and maintained, whenever the potential fall distance exceeds two stories, or 25 feet. The nets are to be hung with sufficient clearance to prevent contacts with the surface of structures below.

Floor periphery-safety railing must be of 1/2-inch wire rope, or its equal, and it is to be installed approximately 42 inches high around the periphery of all temporary-planked floors, or temporary metal-decked floors of tier buildings, and other multi-floored structures, during structural steel assembly.

Where skeleton steel erection is being done, a tightly planked and substantial floor is to be maintained within two stories, or 30 feet, whichever is less, below and directly under that portion of each tier of beams on which any work is being performed, except when gathering and stacking temporary floor planks on a lower floor, in preparation for transferring such planks for use on an upper floor. When gathering and stacking temporary floor planks, the planks are to be removed successively, working toward the last panel of the temporary floor so that the work is always done from the planked floor. Also, when gathering and stacking temporary floor planks from the last panel, the employees assigned to such work are to be protected by safety harnesses with safety lines attached to a catenary line, or other substantial anchorage.

In the erection of a building having double wood floor construction, the rough flooring is to be completed as the building progresses, including the tier below the one on which floor joists are being installed. For single wood floors, or other flooring systems, the floor immediately below the story where the floor joists are being installed, is to be kept planked or decked over.

Structural Steel Assembly (1926.751)

During the final placing of solid web structural members, the load is not to be released from the hoisting line until the members are secured with not less than two bolts, or the equivalent, at each connection, and drawn up wrench tight (see Figure 108).

Figure 108. Steel worker trying to align the members prior to bolting

Open web steel joists are not to be placed on any structural steel framework unless such framework is safely bolted or welded. No load is to be placed on open web steel joists until security requirements are met.

In steel framing, where bar joists are utilized and columns are not framed in at least two directions with structural steel members, a bar joist is to be field-bolted at the columns to provide lateral stability during construction.

Where longspan joists or trusses 40 feet or longer are used, a center row of bolted bridging must be installed in order to provide lateral stability during construction, and this must be done prior to the slacking of the hoisting line. Taglines are to be used for controlling the loads.

Bolting, Riveting, Fitting-Up, and Plumbing-Up (1926.752)

Containers are to be provided for storing or carrying rivets, bolts, and drift pins, and they must be secured against accidental displacement when aloft. When bolts or drift pins are being knocked out, means are to be provided to keep them from falling. If rivet heads are knocked off or backed out, a means shall be provided to keep them from falling. Riveting is not to be done in the vicinity of combustible material unless precautions are taken to prevent fire.

Pneumatic hand tools are to be disconnected from the power source, and the pressure in hose lines is to be released, before any adjustments or repairs are made. Air line hose sections are to be tied together, except when quick disconnect couplers are used to join sections. A safety wire is to be properly installed on the snap and handle of the pneumatic riveting hammer, and it is to be used at all times. The wire size is not to be less than No. 9 (B&S gauge), leaving the handle and annealed No. 14 on the snap, or equivalent. Impact wrenches are to be provided with a locking device for retaining the socket, and eye protection is to be provided.

Connections of the equipment used in plumbing-up are to be properly secured. The turnbuckles are to be secured to prevent unwinding while under stress. Plumbing-up guys, and related equipment are to be placed so that employees can get at the connection points. Plumbing-up guys are to be removed only under the supervision of a competent person.

Wood planking is to be of the proper thickness so it can carry the working load, but it is not to be less than 2 inches thick full size undressed, exterior grade plywood; it must be at least 3/4-inch thick, or equivalent material. Metal decking of sufficient strength is to be laid tight, and secured to prevent movement. Planks must overlap the bearing on each end by a minimum of 12 inches. Wire mesh, exterior plywood, or equivalent, is to be used around columns where planks do not fit tightly. Provisions are to be made to secure temporary flooring against displacement. All unused openings in floors, temporary or permanent, are to be completely planked over or guarded. Workers are to be provided with safety harnesses when they are working on float scaffolds.

TEMPORARY SLEEPING QUARTERS (1926.51)

Any temporary sleeping quarters must be heated, ventilated, and lighted.

TIRE CAGES (1926.600)

Split rims, or rims with locking rings, have been the cause of serious injuries and deaths. Thus, a safety tire rack, cage, or equivalent, to protect workers during tire maintenance and changing, must be provided when inflating, mounting, and dismounting tires on this type of rim.

TOEBOARDS

Generally speaking, toeboards are to be provided when guard rails are required, or when tools and other materials could be accidentally knocked off of a landing, work platform, scaffold, or other raised area. It is also a reminder to workers that they are near the edge. The acceptable height for a toeboard is usually between 3 1/2 inches and 4 inches. The most common height of a toeboard in the construction industry is the height of the width (3 1/2 inches) of a standard 2×4 board (see Figure 109).

Figure 109. 3 1/2 inch toeboard on a guardrail system

TOILETS (1926.51)

Toilets are to be provided according to the number of employees. For 20 or less workers, one toilet is provided; for 20 or more workers, one toilet and one urinal are provided per 40 workers; and for 200 or more workers, one toilet and one urinal are provided per 50 workers. Even under temporary field conditions, at least one toilet is to be available. When no sewer is available, privies, chemical, recirculating, or combustible toilets are permissible (see Figure 110). The previous requirements do not apply to mobile crews who have transportation to a nearby toilet.

Figure 110. Typical toilets found on construction jobsites

TRANSPORTATION

When you have moving vehicles going and coming on a construction site, the potential for errant movement, changes in traffic patterns, and workers getting in the pathway can always occur. All vehicle or heavy equipment operators must obey all traffic signs and posted speeds (See Figure 111). All loads are to be secured properly, and hauled material that overhangs the sides or ends of a truck are to be marked. Workers are never to ride running boards, fenders, siderails, tailgates, or tops of vehicles. Workers are not to extend any part of their body outside of a truck bed, or stand in it while moving. Passengers can only be transported when adequate seats, and/or safety provisions are made. With the movement of vehicular traffic by those who are delivering and removing materials to and from the site, the heavy equipment operators may experience blind spots; therefore, workers themselves must be alert and take precautions to protect themselves from being struck by vehicles and equipment.

Figure 111. Dangers due to loading and movement of truck traffic frequently occur on construction

TUNNELS/SHAFT (UNDERGROUND CONSTRUCTION) (1926.800)

Construction tunnels, and shafts underground, pose the same construction hazards as aboveground. There are also some added problems, the least of which are: a more confining environment with less space, and a lack of stability of the rocks or materials on all sides. Workers must be alert for loose soil, rock, or fractured materials; these should be removed or properly supported. Also, workers are to be alert to moving equipment, especially around loading and hauling equipment.

During underground, construction workers should never work alone. They must be alert to hazards, and get help to correct them. Workers need to wear their personal protective equipment, such as eye protection from dust, flying objects, lasers, or other eye hazards.

Since the underground operations do not have adequate ventilation, a special effort must be made to ensure that the ventilation is adequate at all times. Dust should be controlled during such operations as drilling, and the lighting must be adequate in underground operations.

With space at a premium, waste materials must not be allowed to accumulate in work the areas or passageways. All hoses, lines, and cords from work operations must be protected from damage.

Underground hazards, such as reduced natural ventilation and light, difficult and limited access and egress, exposure to air contaminants, fire, and explosion, can frequently exist. Therefore, in an effort to minimize injuries and illness associated with underground construction, the following safe work procedures are be implemented and enforced during all underground construction operations on company projects.

Underground Construction (1926.800)

The construction of underground tunnels, shafts, chambers, and passageways, as well as cut-and-cover excavations which are both physically connected to ongoing underground construction operations, and covered in such a manner as to create conditions characteristic of underground construction, are all viewed as underground construction.

Egress and Access

The employer's primary responsibility is to provide and maintain a safe means of access and egress to all work stations. This access and egress is to be designed in such a manner that employees are protected from being struck by excavators, haulage machines, trains, and other mobile equipment.

The employer is to control access to all openings in order to prevent unauthorized entry underground. Unused chutes, manways, or other openings are to be tightly covered, bulkheaded, or fenced off, and are to be posted with warning signs indicating "Keep Out" or similar language. Completed or unused sections of the underground facility are to be barricaded.

Check-In/Check-Out

The employer must maintain a check-in/check-out procedure that will ensure that aboveground personnel can determine an accurate count of the number of persons underground in the event of an emergency. However, this procedure is not required when the construction of underground facilities designed for human occupancy has been sufficiently completed so that the permanent environmental controls are effective and no structural failure could occur within the facilities.

All employees are to be instructed in the recognition and avoidance of hazards associated with underground construction activities including, where appropriate, the following subjects: air monitoring, ventilation, illumination, communications, flood control, mechanical equipment, personal protective equipment, explosives, fire prevention and protection, and emergency procedures, including evacuation plans and check-in/check-out systems.

Communications

Any oncoming shifts are to be informed of any hazardous occurrences or conditions that have affected, or might affect employee safety, including liberation of gas, equipment failures, earth or rock slides, cave-ins, floodings, fires, or explosions.

The employer is to establish and maintain direct communications, for the coordination of activities, with other employers whose operations at the jobsite affect, or may affect the safety of employees underground. When natural, unassisted voice communication is ineffective, a power-assisted means of voice communication is to be used to provide communication between the work face, the bottom of the shaft, and the surface. Two effective means of communication, at least one of which is voice communication, are to be provided in all shafts which are being developed, or are used for either personnel access or hoisting.

Powered communication systems must operate on an independent power supply, and are to be installed so that the use of or disruption of any one phone or signal location will not disrupt the operation of the system from any other location. Communication systems are to be tested upon initial entry of each shift to the underground, and as often as necessary at later times, to ensure that they are in working order.

Emergencies

Any employee working alone in a hazardous underground location, who is both out of the range of natural unassisted voice communication, and who is not under observation of other persons, is to be provided with an effective means of obtaining assistance in an emergency.

When a shaft is used as a means of egress, the employer must make advance arrangements for power-assisted hoisting capability to be readily available in an emergency, unless the regular hoisting means can continue to function in the event of an electrical power failure at the jobsite. Such hoisting means are to be designed so that the load hoist drum is powered in both directions of rotation, and so that the brake is automatically applied upon power release or failure.

The employer is to provide self-rescuers, who have current approval from the National Institute for Occupational Safety and Health, and the Mine Safety and Health Administration, to be immediately available to all employees who are at underground work station areas and might be trapped by smoke or gas.

At least one designated person needs to be on duty, aboveground, whenever any employee is working underground.   This designated person is to be responsible for securing immediate aid, and must keep an accurate count of the employees underground, in case of emergency. The designated person must not be so busy that the counting function is encumbered.

Each employee underground shall have an acceptable portable hand lamp, or cap lamp, in his or her work area for emergency use, unless natural light or an emergency lighting system provides adequate illumination for escape.

On jobsites where 25 or more employees work underground at one time, the employer must provide (or make arrangements in advance with locally available rescue services to provide) at least two 5-person rescue teams – one on the jobsite, or within one-half hour travel time from the entry point, and the other one within a 2 hour travel time. On jobsites where less than 25 employees work underground at one time, the employer must provide (or make arrangements in advance with locally available rescue services to provide) at least one 5-person rescue team to be either on the jobsite, or within one-half hour travel time from the entry point.

Rescue team members are to be qualified in rescue procedures, the use and limitations of breathing apparatus, and the use of firefighting equipment. Qualifications shall be reviewed not less than annually. On jobsites, where flammable or noxious gases are encountered or anticipated in hazardous quantities, rescue team members must practice donning and using self-contained breathing apparatus, monthly. The employer must ensure that rescue teams are familiar with conditions at the jobsite.

Gassy Operations

Underground construction operations are to be classified as potentially gassy: if either air monitoring discloses 10 percent or more of the lower explosive limit for methane; if other flammable gases measured at 12 inches to + or – 0.25 inch from the roof, face, floor, or walls, in any underground work area, for more than a 24-hour period on three consecutive days; if the history of the geographical area or geological formation indicates that 10 percent or more of the lower explosive limit for methane, or other flammable gases are likely to be encountered in such underground operations; if there has been an ignition of methane, or of other flammable gases emanating from the strata which indicates the presence of such gases; and if the underground construction operations are both connected to an underground work area, which is currently classified as gassy, and is also subject to a continuous course of air containing the flammable gas concentration.

Underground construction gassy operations may be declassified to Potentially Gassy, when air monitoring results remain under 10 percent of the lower explosive limit for methane or other flammable gases, for three consecutive days. Only acceptable equipment, maintained in suitable condition, shall be used in gassy operations. Mobile diesel-powered equipment, used in gassy operations, is to be approved or verified to be equivalent to requirements of 30 CFR Part 36. Each entrance to a gassy operation is to be prominently posted with signs notifying all entrants of the gassy classification, and smoking is to be prohibited in all gassy operations. The employer is responsible for collecting all personal sources of ignition. A fire watch is to be maintained when hot work is performed.

Once an operation has been classified as gassy, all operations in the affected area, except the following, are to be discontinued until the operation is either in compliance with all of the gassy operation requirements, or has been declassified.

1.   Operations related to the control of the gas concentration.

2.   Installation of new equipment, or conversion of existing equipment to comply with requirements.

3.   Installation of above-ground controls for reversing the air flow.

Air Quality

The employer must assign a competent person to perform all air monitoring. The competent person shall make a reasonable determination as to which substances to monitor, and how frequently to monitor. This is to be based on the location of the jobsite, including the proximity to fuel tanks, sewers, gas lines, old landfills, coal deposits, swamps, and geology of the jobsite, particularly those involving the soil type and its permeability presence of air contaminants in nearby jobsites, as well as the changes in the levels of substances which were monitored on the prior shift. Also the following work practices and jobsite conditions should be taken into account: the use of diesel engines, the use of explosives, the use of fuel gas, the volume and flow of ventilation, the visible atmospheric conditions, the decompression of the atmosphere, welding, cutting and hot work, and the employees' physical reactions to working underground.

The atmosphere in all underground work areas is to be tested, as often as necessary, to ensure that the atmosphere, at normal atmospheric pressure, contains at least 19.5 percent oxygen, and no more than 22 percent oxygen. The atmosphere in all underground work areas is to be tested quantitatively for carbon monoxide, nitrogen dioxide, hydrogen sulfide, and other toxic gases, dusts, vapors, mists, and fumes, as often as necessary, to ensure that the permissible exposure limits prescribed are not exceeded. The atmosphere in all underground work areas is to be tested quantitatively for methane and flammable gases as often as necessary.

If diesel-engine, or gasoline-engine driven ventilating fans or compressors are used, an initial test is to be made of the inlet air of the fan or compressor, with the engines operating, to ensure that the air supply is not contaminated by engine exhaust.

When rapid excavation machines are used, a continuous flammable gas monitor is to be operated at the face, with the sensor(s) placed as high and close to the front of the machine's cutter head as practicable.

Hydrogen Sulfide

Whenever air monitoring indicates the presence of 5 ppm or more of hydrogen sulfide, a test is to be conducted in the affected underground work area(s), at least at the beginning and midpoint of each shift, until the concentration of hydrogen sulfide has been less than 5 ppm for 3 consecutive days. Whenever hydrogen sulfide is detected in an amount exceeding 10 ppm, a continuous sampling and indicating hydrogen sulfide monitor is to be used to monitor the affected work area. Employees are to be informed when a concentration of 10 ppm hydrogen sulfide is exceeded. The continuous sampling and indicating hydrogen sulfide monitor is to be designed, installed, and maintained to provide a visual and aural alarm, when the hydrogen sulfide concentration reaches 20 ppm, to signal that additional measures, such as respirator use, increased ventilation, or evacuation, might be necessary to maintain hydrogen sulfide exposure below the permissible exposure limit.

When the competent person determines, on the basis of air monitoring results or other information, that air contaminants may be present in sufficient quantity to be dangerous to life, the employer must prominently post a notice at all entrances to the underground jobsite to inform all entrants of the hazardous condition, and must ensure that the necessary precautions are taken.

Flammable Gases

Whenever five percent or more of the lower explosive limit for methane or other flammable gases is detected in any underground work area(s), or in the air return, steps are to be taken to increase ventilation air volume, or to otherwise control the gas concentration, unless the employer is operating in accordance with the potentially gassy, or gassy operation requirements. Such additional ventilation controls may be discontinued when gas concentrations are reduced below five percent of the lower explosive limit, but are reinstituted whenever the five percent level is exceeded. Whenever 10 percent or more of the lower explosive limit for methane or other flammable gases is detected in the vicinity of welding, cutting, or other hot work, such work is to be suspended until the concentration of such flammable gas is reduced to less than 10 percent of the lower explosive limit.

Whenever 20 percent or more of the lower explosive limit for methane or other flammable gases is detected in any underground work area(s), or in the air return, all employees, except those necessary to eliminate the hazard, are to be immediately withdrawn to a safe location aboveground. Electrical power, except for acceptable pumping and ventilation equipment, is to be cut off to the area endangered by the flammable gas, until the concentration of such gas is reduced to less than 20 percent of the lower explosive limit.

Operations which meet the criteria for potentially gassy and gassy operations are subject to the additional monitoring. A test for oxygen content is to be conducted in the affected underground work areas, and the work areas immediately adjacent to such areas, at least at the beginning and midpoint of each shift. When using rapid excavation machines, continuous automatic flammable gas monitoring equipment is to be used to monitor the air at the heading, on the rib, and in the return air duct. The continuous monitor signals the heading and shuts down electric power in the affected underground work area, except for acceptable pumping and ventilation equipment, when 20 percent or more of the lower explosive limit for methane or other flammable gases are encountered. A manual flammable gas monitor is to be used, as needed, and at least at the beginning and midpoint of each shift, to ensure that the limits are not exceeded. In addition, a manual electrical shutdown control is to be provided near the heading. Local gas tests are to be made prior to, and continuously during any welding, cutting, or other hot work.

In underground operations driven by drill-and-blast methods, the air in the affected area is to be tested, after blasting, and prior to re-entry, for flammable gas, and it is to be tested continuously when employees are working underground.

A record of all air quality tests is to be maintained aboveground at the worksite, and it is to be made available upon request. The record must include the location, date, time, substance, and amount monitored. Records of exposures to toxic substances are to be retained. All other air quality test records shall be retained until completion of the project.

Ventilation

Fresh air is to be supplied to all underground work areas in sufficient quantities to prevent dangerous or harmful accumulation of dusts, fumes, mists, vapors, or gases. Mechanical ventilation is to be provided in all underground work areas, except when the employer can demonstrate that natural ventilation provides the necessary air quality through sufficient air volume and air flow. A minimum of 200 cubic feet of fresh air per minute is to be supplied for each employee underground. The linear velocity of air flow in the tunnel bore, in shafts, and in all other underground work areas, must be at least 30 feet per minute where blasting or rock drilling is conducted, or where other conditions which are likely to produce dust, fumes, mists, vapors, or gases in harmful or explosive quantities, are present. The direction of the mechanical air flow must be reversible. Following blasting, ventilation systems should exhaust smoke and fumes to the outside atmosphere before work is resumed in affected areas.

Ventilation doors are to be designed and installed so that they remain closed when in use, regardless of the direction of the air flow. When ventilation has been reduced to the extent that hazardous levels of methane or flammable gas may have accumulated, a competent person must test all affected areas after ventilation has been restored, and the competent person must determine whether the atmosphere is within flammable limits, before any power, other than for acceptable equipment, is restored, or before work is resumed. Whenever the ventilation system has been shut down with all employees out of the underground area, only competent persons, who are authorized to test for air contaminants, are allowed to be underground until the ventilation has been restored and all affected areas have been tested for air contaminants and declared safe.

When drilling rocks or concrete, appropriate dust control measures are to be taken to maintain dust levels within limits. Such measures may include, but are not limited to, wet drilling, the use of vacuum collectors, and water mix spray systems. Internal combustion engines, except diesel-powered engines on mobile equipment, are prohibited underground. Mobile diesel-powered equipment, used underground in atmospheres other than gassy operations, is either to be approved or certified by the employer as equivalent. (Each brake horsepower of a diesel engine requires at least 100 cubic feet of air per minute for suitable operation in addition to the air requirements for personnel. Some engines may require a greater amount of air to ensure that the allowable levels of carbon monoxide, nitric oxide, and nitrogen dioxide are not exceeded.)

Potentially gassy or gassy operations need to have ventilation systems installed which are constructed of fire-resistant materials and must have acceptable electrical systems, including fan motors.

Gassy operations are to be provided with controls, located aboveground, for reversing the air flow of ventilation systems. In potentially gassy or gassy operations, where mine-type ventilation systems are used which have an offset main fan installed on the surface, they are to be equipped with explosion-doors, or have a weak-wall which has an area at least equivalent to the cross-sectional area of the airway.

Illumination

Illumination requirements, applicable to underground construction operations, must be employed. Only acceptable portable lighting equipment is to be used within 50 feet of any underground heading during explosives handling.

Fire Prevention

Fire prevention and protection requirements, applicable to underground construction operations, are to be followed. Open flames and fires are to be prohibited in all underground construction operations, except as permitted for welding, cutting, and other hot work operations. Smoking may be allowed only in areas free of fire and explosion hazards. Readily visible signs prohibiting smoking and open flames are to be posted in areas having fire explosion hazards. The employer is not permitted to store underground more than a 24-hour supply of diesel fuel for the underground equipment used at the worksite. The piping of diesel fuel, from the surface to an underground location, is permitted only if the diesel fuel is contained at the surface. This fuel must be in a tank which has a maximum capacity of no more than the amount of fuel required to supply, for a 24-hour period, the equipment serviced by the underground fueling station. The surface tank is to be connected to the underground fueling station by an acceptable pipe or hose system, and this system is to be controlled at the surface by a valve, and at the shaft bottom by a hose nozzle. The pipe is to be empty at all times, except when transferring diesel fuel from the surface tank to a piece of equipment in use underground. In the shaft, hoisting operations are to be suspended during refueling operations, if the supply piping, in the shaft, is not protected from damage. Gasoline is not to be carried, stored, or used underground. Acetylene, liquefied petroleum gas, and methylacetylene propadiene stabilized gas may be used underground only for welding, cutting, and other hot work.

Oil, grease, and diesel fuel, stored underground, are to be kept in tightly sealed containers and in fire-resistant areas at least 300 feet from underground explosive magazines. They are to be at least 100 feet from shaft stations and steeply inclined passageways. Storage areas are to be positioned or diked so that the contents of ruptured or overturned containers will not flow from the storage area.

Flammable or combustible materials are not to be stored aboveground within 100 feet of any access opening to any underground operation. Where this is not feasible because of space limitations at the jobsite, such materials may be located within the 100-foot limit, provided that they are located as far as practicable from the opening. Either a fire-resistant barrier, of not less than one-hour rating, is to be placed between the stored material and the opening, or additional precautions are to be taken which will protect the materials from ignition sources. Fire-resistant hydraulic fluids are to be used in hydraulically actuated underground machinery and equipment unless such equipment is protected by a fire suppression system, or by multipurpose fire extinguisher(s), rated at sufficient capacity for the type and size of hydraulic equipment involved, but rated at least 4A:40B:C.

Electrical installations in underground areas where oil, grease, or diesel fuel are stored, are to be used only for lighting fixtures. Lighting fixtures in storage areas, or within 25 feet of underground areas where oil, grease, or diesel fuel are stored, are to be approved for Class I, Division 2 locations.

Leaks and spills of flammable or combustible fluids are to be cleaned up immediately. A fire extinguisher of at least 4A:40B:C rating, or other equivalent extinguishing means, is to be provided at the head pulley, and at the tail pulley of underground belt conveyers. Any structure located underground, or within 100 feet of an opening to the underground, is to be constructed of material having a fire-resistance rating of at least one hour.

When underground welding, cutting, and other hot work is occurring, no more than the amount of fuel gas and oxygen cylinders necessary to perform welding, cutting, or other hot work during the next 24-hour period, is permitted underground, and noncombustible barriers are to be installed below welding, cutting, or other hot work being done in, or over a shaft or raise.

Unstable Formations

Portal openings and access areas are to be guarded by shoring, fencing, head walls, shotcreting, or other equivalent protection, to ensure safe access of employees and equipment. Adjacent areas are to be scaled, or otherwise secured to prevent loose soil, rock, or fractured materials from endangering the portal and access area. The employer must ensure ground stability in hazardous subsidence areas by shoring, by filling in, or by erecting barricades and posting warning signs to prevent entry. A competent person is to inspect the roof, face, and walls of the work area at the start of each shift, and as often as necessary to determine ground stability. Competent persons, who conduct such inspections, must be protected from loose ground by location, by ground support, or by equivalent means. Ground conditions along haul-ageways and travelways are to be inspected as frequently as necessary to ensure safe passage. Loose ground that might be hazardous to employees is to be taken down, scaled, or supported.

Torque wrenches are to be used wherever bolts, used for ground support, depend on torsionally applied force. A competent person must determine whether rock bolts meet the necessary torque, and must also determine the testing frequency in light of the bolt system, ground conditions, and the distance from vibration sources.

Suitable protection is to be provided for employees exposed to the hazard of loose ground while installing ground support systems. Support sets are to be installed so that the bottoms have sufficient anchorage to prevent ground pressures from dislodging the support base of the sets. Lateral bracing (collar bracing, tie rods, or spreaders) is to be provided between immediately adjacent sets in order to ensure added stability. Damaged or dislodged ground supports that create a hazardous condition are to be promptly repaired or replaced. When replacing supports, the new supports are to be installed before the damaged supports are removed.

A shield, or other type of support, is to be used to maintain a safe travelway for employees who are working in dead-end areas and are ahead of any support replacement operation. Where employees must enter shafts and wells over 5 feet in depth, the shafts and wells are to be supported by a steel casing, concrete pipe, timber, solid rock, or other suitable material.

The full depth of the shaft is to be supported by casing, or bracing, except where the shaft penetrates into solid rock having characteristics that will not change as a result of exposure. Where the shaft passes through earth into solid rock, or through solid rock into earth, and where there is a potential for shear, the casing or bracing must extend at least 5 feet into the solid rock. When the shaft terminates in solid rock, the casing or bracing must extend to the end of the shaft, or 5 feet into the solid rock, whichever is less. The casing or bracing must extend 42 inches, plus or minus 3 inches, aboveground level, except that the minimum casing height may be reduced to 12 inches, provided that a standard railing is installed; that the ground adjacent to the top of the shaft is sloped away from the shaft collar to prevent entry of liquids; and that effective barriers are used to prevent mobile equipment operating near the shaft from jumping over the 12 inch barrier.

Explosives and Blasting

No explosives or blasting agents are to be permanently stored in any underground operation until the operation has been developed to the point where at least two modes of exit have been provided. Permanent underground storage magazines are to be at least 300 feet from any shaft, adit, or active underground working area. Permanent underground magazines, containing detonators, are not to be located closer than 50 feet to any magazine containing other explosives or blasting agents.

When using and transporting explosives while underground, special precautions are to be taken. All explosives or blasting agents in transit underground are to be taken to the place of use, or storage, without delay. The quantity of explosives or blasting agents taken to an underground loading area must not exceed the amount estimated to be necessary for the blast. Again, explosives in transit are not to be left unattended. The hoist operator is to be notified before explosives or blasting agents are transported in a shaft conveyance. Trucks used for the transportation of explosives underground must have the electrical system checked weekly to detect any failures which may constitute an electrical hazard. A certification record, which includes the date of the inspection, the signature of the person who performed the inspection, and a serial number, or other identifier of the truck inspected, is to be prepared, and the most recent certification record is to be maintained on file. The installation of auxiliary lights on truck beds, which are powered by the truck's electrical system, is prohibited.

When firing from a power circuit in underground operations, a safety switch is to be placed, at intervals, in the permanent firing line. This switch is to be made so it can be locked only in the "Off" position, and it is to be provided with a short-circuiting arrangement of the firing lines to the cap circuit. In underground operations, there is to be a "lightning" gap of at least 5 feet in the firing system, ahead of the main firing switch that is between this switch and the source of power. This gap is to be bridged by a flexible jumper cord just before firing the blast.

During underground operations when explosives and blasting agents are hoisted, lowered, or conveyed in a powder car, no other materials, supplies, or equipment are to be transported in the same conveyance, at the same time. No one, except the operator, his helper, and the powderman, is permitted to ride on a conveyance transporting explosives and blasting agents. No person is to ride in any shaft conveyance transporting explosives and blasting agents. No explosives or blasting agents are to be transported on any locomotive. At least two car lengths must separate the locomotive from the powder car. No explosives or blasting agents are to be transported on a man haul trip. The car or conveyance containing explosives or blasting agents is to be pulled, not pushed, whenever possible. The powder car, or conveyance, should be especially built for the purpose of transporting explosives or blasting agents. It must bear a reflectorized sign on each side, with the word "Explosives" in letters not less than 4 inches in height, and these letters are to be placed upon a background of sharply contrasting color. Compartments, for transporting detonators and explosives in the same car or conveyance, are to be physically separated by a distance of 24 inches, or by a solid partition at least 6 inches thick. Detonators, and other explosives, are not to be transported at the same time in any shaft conveyance. Explosives, blasting agents, or blasting supplies are not to be transported with other materials. Explosives or blasting agents, not in original containers, are to be placed in a suitable container when transported manually. Detonators, primers, and other explosives are to be carried in separate containers when transported manually.

When blasting under compressed air during excavation work, detonators and explosives are not to be stored or kept in tunnels, shafts, or caissons. Detonators and explosives, for each round, are to be taken directly from the magazines to the blasting zone, and immediately loaded. Detonators and explosives left over after loading a round are to be removed from the working chamber before the connecting wires are connected up. When detonators or explosives are brought into an air lock, no employee, except the powderman, blaster, lock tender, and the employees necessary for carrying, are permitted to enter the air lock. No other material, supplies, or equipment are to be locked through with the explosives. Detonators and explosives are to be taken separately into pressure working chambers. The blaster or powderman is responsible for the receipt, unloading, storage, and on-site transportation of explosives and detonators. All metal pipes, rails, air locks, and steel tunnel lining are to be electrically bonded together and grounded at or near the portal or shaft, and such pipes and rails are to be cross-bonded together at not less than 1,000-foot intervals throughout the length of the tunnel. In addition, each low air supply pipe shall be grounded at its delivery end. No explosives are to be loaded or used underground in the presence of combustible gases or combustible dusts.

The explosives suitable for use in wet holes are to be water resistant and shall be Fume Class 1. When tunnel excavation in rock face is approaching a mixed face, or when tunnel excavation is in a mixed face, blasting is to be performed with light charges and with light burden on each hole. Advance drilling is to be performed, as tunnel excavation in rock face approaches mixed face, to determine the general nature and extent of the rock cover, and the remaining distance ahead to soft ground as the excavation advances.

After blasting operations in shafts, a competent person must determine if the walls, ladders, timbers, blocking, or wedges have loosened. If so, necessary repairs are to be made before employees, other than those assigned to make the repairs, are allowed in or below the affected areas. Further requirements for blasting and explosives operations, including handling of misfires, are found in 29 CFR 1926 Subpart U.

Blasting wires are to be kept clear of electrical lines, pipes, rails, and other conductive material, excluding earth, to prevent explosives' initiation, or employee exposure to electric current. Following blasting, workers should not enter a work area until the air quality meets all the requirements.

Drilling

A competent person must inspect all drilling and associated equipment prior to each use. Equipment defects, which affect safety, are to be corrected before the equipment is used. The drilling area is to be inspected for hazards before the drilling operation is started. Employees are not allowed on a drill mast while the drill bit is in operation, or the drill machine is being moved. When a drill machine is being moved from one drilling area to another, drill steel, tools, and other equipment must be secured and the mast is to be placed in a safe position. Receptacles or racks need to be provided for storing drill steel located on jumbos.

Employees working below jumbo decks are to be warned whenever drilling is about to begin. Drills on columns are to be anchored firmly before starting drilling, and are to be retightened, as necessary, thereafter. The employer must provide a mechanical means on the top deck of a jumbo for lifting unwieldy or heavy material. When jumbo decks are over 10 feet in height, the employer shall install stairs wide enough for two persons. Jumbo decks more than 10 feet in height are to be equipped with guardrails on all open sides, excluding access openings of platforms, unless an adjacent surface provides equivalent fall protection.

Only employees assisting the operator are to be allowed to ride on jumbos, unless the jumbo meets the specific requirements. Jumbos are to be chocked to prevent movement while employees are working on them. Walking and working surfaces of jumbos are to be maintained to prevent the hazards of slipping, tripping, and falling. Jumbo decks and stair treads are to be designed to be slip resistant, and they are to be secured to prevent accidental displacement.

General Guidelines

Scaling bars are to be available at scaling operations, and are to be maintained in good condition at all times. Blunted, or severely worn bars, are not to be used. Blasting holes are not to be drilled through blasted rock (muck) or water. Employees in a shaft are to be protected either by location, or by suitable barrier(s), if powered mechanical loading equipment is used to remove muck which contains unfired explosives. A caution sign which reads, "Buried Line," or similar wording, is to be posted where air lines are buried, or otherwise hidden by water or debris.

Power Haulage

A competent person is to inspect haulage equipment before each shift. Equipment defects which affect safety and health are to be corrected before the equipment is used. Powered mobile haulage equipment must have a suitable means of stopping. Power mobile haulage equipment, including trains, are to be equipped with audible warning devices to warn employees to stay clear.

The operator must sound the warning device before moving the equipment, and whenever necessary during travel. The operator must assure that lights, which are visible to employees at both ends of any mobile equipment, including a train, are turned on whenever the equipment is operating. In those cabs where glazing is used, the glass is to be safety glass, or its equivalent, and is to be maintained and cleaned so that vision is not obstructed.

Anti-roll back devices or brakes are to be installed on inclined conveyer drive units to prevent conveyers from inadvertently running in reverse. Employees are not permitted to ride a power-driven chain, belt, or bucket conveyer unless the conveyer is specifically designed for the transportation of persons. Endless belt type manlifts are prohibited in underground construction. The usual rules for conveyors apply.

No employee shall ride haulage equipment unless it is equipped with seating for each passenger, and it protects passengers from being struck, crushed, or caught between other equipment or surfaces. Members of train crews may ride on a locomotive, if it is equipped with handholds, and nonslip steps or footboards.

Powered mobile haulage equipment, including trains, are not to be left unattended unless the master switch or motor is turned off. The operating controls are to be in neutral or the park position, and the brakes are to be set, or equivalent precautions are to be taken to prevent rolling. Whenever rails serve as a return for a trolley circuit, both rails are to be bonded at every joint, and crossbonded every 200 feet.

When dumping cars by hand, the car dumps must have tiedown chains, bumper blocks, or other locking or holding devices to prevent the cars from overturning. Rocker-bottom or bottom-dump cars are to be equipped with positive locking devices to prevent the cars from overturning.

Equipment to be hauled is to be loaded and secured to prevent sliding or dislodgment. Mobile equipment, including rail-mounted equipment, is to be stopped for manual connecting or service work. Employees are not to reach between moving cars during coupling operations. Couplings are not to be aligned, shifted, or cleaned on moving cars or locomotives. Safety chains or other connections are to be used, in addition to couplers, to connect man cars, or powder cars, whenever the locomotive is uphill of the cars.

When the grade exceeds one percent and there is a potential for runaway cars, safety chains or other connections are to be used, in addition to couplers, to connect haulage cars or, as an alternative, the locomotive must be downhill of the train. Such safety chains or other connections shall be capable of maintaining connections between cars in the event of coupler disconnect, failure, or breakage. Parked rail equipment is to be chocked, blocked, or have the brakes act to prevent inadvertent movement.

Berms, bumper blocks, safety hooks, or equivalent means, are to be provided to prevent overtravel and overturning of haulage equipment at dumping locations. Bumper blocks, or equivalent stopping devices, are to be provided at all track dead ends.

Only small handtools, lunch pails, or similar small items may be transported with employees in mancars, or on top of a locomotive. When small hand tools or other small items are carried on top of a locomotive, the top is to be designed or modified to retain them while traveling.

Where switching facilities are available, occupied personnel-cars are to be pulled, not pushed. If personnel-cars must be pushed and visibility of the track ahead is hampered, a qualified person is to be stationed in the lead car to give signals to the locomotive operator. Crew trips are to consist of personnel-loads only.

Electrical Safety

In addition to the normal electrical construction safety requirements, electric power lines are to be insulated, or located away from water lines, telephone lines, air lines, or other conductive materials so that a damaged circuit will not energize the other systems. Lighting circuits are to be located so that movement of personnel or equipment will not damage the circuits or disrupt service. Oil-filled transformers are not to be used underground, unless they are located in a fire-resistant enclosure which is suitably vented to the outside, and surrounded by a dike to retain the contents of the transformers in the event of rupture.

Cranes

Although the usual crane requirements apply, there are other general requirements for underground cranes and hoists which deal with materials, tools, and supplies that are being raised or lowered. Whether these materials, tools, or supplies are within a cage or otherwise, they are to be secured or stacked in a manner to prevent the load from shifting, snagging, or falling into the shaft. A warning light, suitably located, is to flash and warn employees at the shaft bottom, and subsurface shaft entrances, whenever a load is above the shaft bottom or subsurface entrances, or the load is being moved in the shaft. This does not apply to fully enclosed hoistways.

Whenever a hoistway is not fully enclosed and employees are at the shaft bottom, conveyances or equipment are to be stopped at least 15 feet above the bottom of the shaft and held there until the signalman at the bottom of the shaft directs the operator to continue lowering the load, except that the load may be lowered without stopping, if the load or conveyance is within full view of a bottom signalman who is in constant voice communication with the operator.

Before maintenance, repairs, or other work commence in a shaft served by a cage, skip, or bucket, the operator and other employees in the area are to be informed and given suitable instructions. A sign warning that work is being done in the shaft is to be installed at the shaft collar, at the operator's station, and at each underground landing.

Any connection between the hoisting rope and the cage or skip is to be compatible with the type of wire rope used for hoisting. Spin-type connections, where used, are to be maintained in a clean condition, and protected from foreign matter that could affect their operation. Cage, skip, and load connections to the hoist rope are to be made so that the force of the hoist pull, vibration, misalignment, release of lift force, or impact will not disengage the connection. Moused or latched open-throat hooks do not meet this requirement. When using wire rope wedge sockets, means are to be provided to prevent wedge escapement, and to ensure that the wedge is properly seated.

Cranes are to be equipped with a limit switch to prevent overtravel at the boom tip. Limit switches are to be used only to limit travel of loads when operational controls malfunction, and are not used as a substitute for other operational controls. Hoists are to be designed so that the load hoist drum is powered in both directions of rotation, and so that brakes are auto-matically applied upon power release or failure.

Control levers are to be of the "deadman type" which return automatically to their center (neutral) position upon release. When a hoist is used for both personnel hoisting and material hoisting, load and speed ratings for personnel and for materials are to be assigned to the equipment. Material hoisting may be performed at speeds higher than the rated speed for personnel hoisting, if the hoist and components have been designed for such higher speeds, and if shaft conditions permit.

Employees do not ride on top of any cage, skip, or bucket, except when necessary to perform inspection or maintenance of the hoisting system, in which case they are to be pro-tected by a body harness system to prevent falling.

Personnel and materials (other than small tools and supplies which are secured in a manner that will not create a hazard to employees) are not to be hoisted together in the same conveyance. However, if the operator is protected from the shifting of materials, then the operator may ride with materials in cages or skips which are designed to be controlled by an operator within the cage or skip.

Line speed must not exceed the design limitations of the systems. Hoists are to be equipped with landing level indicators at the operator's station. Marking the hoist rope does not satisfy this requirement.

Whenever glazing is used in the hoist house, it is to be safety glass, or its equivalent, and must be free of distortions and obstructions. A fire extinguisher that is rated at least 2A:10B:C (multi-purpose, dry chemical) is to be mounted in each hoist house. Hoist controls are to be arranged so that the operator can perform all operating cycle functions, and reach the emergency power cutoff without having to reach beyond the operator's normal operating position.

Hoists are to be equipped with limit switches to prevent overtravel at the top and bottom of the hoistway. Limit switches are to be used only to limit travel of loads when opera-tional controls malfunction, and they are not to be used as a substitute for other operational controls. Hoist operators are to be provided with a closed-circuit voice communication system to each landing station, with speaker microphones so located that the operator can communi-cate with individual landing stations during hoist use.

When sinking shafts 75 feet or less in depth, cages, skips, and buckets that may swing, bump, or snag against shaft sides or other structural protrusions, are to be guided by fenders, rails, ropes, or a combination of those means. When sinking shafts more than 75 feet in depth, all cages, skips, and buckets are to be rope or rail guided to within a rail length from the sinking operation. Cages, skips, and buckets in all completed shafts, or in all shafts being used as completed shafts, are to be rope or rail-guided for the full length of their travel.

Wire rope used in load lines of material hoists is to be capable of supporting, without failure, at least five times the maximum intended load, or the factor recommended by the rope manufacturer, whichever is greater. The design factor is to be calculated by dividing the break-ing strength of wire rope, as reported in the manufacturer's rating tables, by the total static load, including the weight of the wire rope in the shaft when fully extended.

A competent person is to visually check all hoisting machinery, equipment, anchor-ages, and hoisting rope at the beginning of each shift, and during hoist use, as necessary. Each safety device is to be checked by a competent person, at least weekly during hoist use, to ensure suitable operation and safe condition.

In order to ensure suitable operation and the safe condition of all functions and safety devices, each hoist assembly is to be inspected and load-tested to 100 percent of its rated capacity at the time of installation; after any repairs or alterations affecting its structural integ-rity; after the operation of any safety device; and annually when in use. The employer must prepare a certification record, which includes the date each inspection and load-test was performed; the signature of the person who performed the inspection and test; and a serial number or other identifier for the hoist that was inspected and tested. The most recent certification record is to be maintained on file until completion of the project.

Before hoisting personnel or material, the operator must perform a test run of any cage or skip, whenever it has been out of service for one complete shift, and whenever the assembly or components have been repaired or adjusted. Unsafe conditions are to be corrected before using the equipment.

Hoist drum systems are to be equipped with at least two means of stopping the load, each of which are capable of stopping and holding 150 percent of the hoist's rated line pull. A broken-rope safety, safety catch, or arrestment device is not a permissible means for stopping. The operation must remain within sight and sound of the signals at the operator's station.

All sides of personnel cages are to be enclosed by one-half inch wire mesh (not less than No. 14 gauge or equivalent) to a height of not less than 6 feet. However, when the cage or skip is being used as a work platform, its sides may be reduced in height to 42 inches, when the conveyance is not in motion.

All personnel cages are to be provided with positive locking doors that do not open outward. All personnel cages are to be provided with a protective canopy. The canopy is to be made of steel plate, at least 8/16-inch in thickness, or be material of equivalent strength and impact resistance. The canopy is to be sloped to the outside, and be so designed that a section may be readily pushed upward to afford emergency egress. The canopy must cover the top in such a manner as to protect those inside from objects falling in the shaft.

Personnel platforms, operating on guide rails or guide ropes, are to be equipped with broken-rope safety devices, safety catches, or arrestment devices that will stop and hold 150 percent of the weight of the personnel platform and its maximum rated load. During sinking operations in shafts where guides and safeties are not yet used, the travel speed of the personnel platform must not exceed 200 feet per minute. Governor controls set for 200 feet per minute are to be installed in the control system and shall be used during personnel hoisting. The personnel platform may travel over the controlled length of the hoistway at rated speeds up to 600 feet per minute, during sinking operations, in shafts where guides and safeties are used. The personnel platform may travel at rated speeds greater than 600 feet per minute in completed shafts.

Caissons (1926.801)

In caisson work in which compressed air is used, and the working chamber is less than 11 feet in length, and when such caissons are at any time suspended or hung while work is in progress, so that the bottom of the excavation is more than 9 feet below the deck of the working chamber, a shield is to be erected therein for the protection of the employees. Shafts are to be subjected to a hydrostatic or air-pressure test, at the pressure at which they shall be air tight. The shaft is to be stamped on the outside shell, about 12 inches from each flange, to show the pressure to which they have been subjected. Whenever a shaft is used, it is to be provided, where space permits, with a safe, proper, and suitable staircase for its entire length, including landing platforms, not more than 20 feet apart. Where this is impracticable, suitable ladders are to be installed, with landing platforms located about 20 feet apart to break the climb.

All caissons, having a diameter or side greater than 10 feet, are to be provided with a man lock and shaft for the exclusive use of employees. In addition to the gauge in the locks, an accurate gauge is to be maintained on the outer and inner side of each bulkhead. These gauges are to be accessible at all times, and are to be kept in accurate working order. In caisson operations, where employees are exposed to compressed air working environments, standard compressed air guidelines are to be followed.

Cofferdams (1926.802)

If overtopping of the cofferdam by high waters is possible, means are to be provided for controlled flooding of the work area. Warning signals, for the evacuation of employees in case of emergency, are to be developed and posted. Cofferdam walkways, bridges, or ramps with at least two means of rapid exit, are to be provided with guardrails. Cofferdams located close to navigable shipping channels are to be protected from vessels in transit, where possible.

Compressed Air (1926.803)

When work is occurring under compressed air, there must be present, at all times, at least one competent person, designated by and representing the employer, who is familiar with all aspects, and responsible for full compliance of working under compressed air.

Medical Requirements

Every employee is to be instructed in the rules and regulations which concern their safety, or the safety of others. There must be at least one or more licensed physicians retained who is familiar with, and experienced in the physical requirements and the medical aspects of compressed air work, and the treatment of decompression illness. The physician must be available at all times while work is in progress, in order to provide medical supervision for employees employed in compressed air work. The physician must be physically qualified, and must be willing to enter a pressurized environment.

No employee is to be permitted to enter a compressed air environment until examined by the physician, and reported to be physically qualified to engage in such work. In the event an employee is absent from work for 10 days, or is absent due to sickness or injury, that employee shall not resume work until reexamined by the physician. After the examination, the employee's physical condition is to be reported, by the physician, to be such as to permit the worker to work in compressed air. After an employee has been continuously employed in compressed air work for a period, designated by the physician, but not to exceed one year, the worker is to be reexamined by the physician to determine if the employee is still physically qualified to engage in compressed air work.

Such physicians must, at all times, keep a complete and full record of the examinations made, and must keep accurate records of any decompression illness, or other illness or injury, which incapacitates any employee for work. The physicians must also report all loss of life that occurs in the operation of a tunnel, caisson, or other compartment in which compressed air is used. Records are to be available for the inspection of the Secretary or his representatives, and a copy is to be forwarded to OSHA within 48 hours following the occurrence of the accident, death, injury, or decompression illness. It is to state, as fully as possible, the cause of said death or decompression illness, the place where the injured or sick employee was taken, and such other relative information as may be required by the Secretary.

A fully equipped first aid station is provided at each tunnel project regardless of the number of persons employed. An ambulance or transportation suitable for a litter case is to be at each project. Where tunnels are being excavated from portals more than 5 road miles apart, a first aid station and transportation facilities are to be provided at each portal.

Medical Lock

A medical lock is to be established and maintained in the immediate working order, whenever air pressure in the working chamber is increased above the normal atmosphere. The medical lock has to have at least 6 feet of clear headroom at the center, be subdivided into not less than two compartments, and is to be readily accessible to employees working under compressed air. The medical lock is to be kept ready for immediate use for at least 5 hours subsequent to the emergence of any employee from the working chamber. It is to be properly heated, lighted, ventilated, and maintained in a sanitary condition. A non-shatterable port shall be present through which the occupant(s) may be kept under constant observation. The medical lock is to be designed for a working pressure of 75 psig.; equipped with internal controls which may be overridden by external controls; provided with air pressure gauges to show the air pressure within each compartment to observers inside and outside the medical lock; equipped with a manual type sprinkler system that can be activated inside the lock, or by the outside lock tender; and provided with oxygen lines and fittings leading into external tanks. The lines are to be fitted with check valves to prevent reverse flow. The oxygen system inside the chamber is to be of a closed circuit design, and to be so designed as to automatically shut off the oxygen supply whenever the fire system is activated. An attendant is to be in constant charge, and under the direct control of the retained physician. The attendant is to be trained in the use of the lock; to be suitably instructed regarding the steps to be taken in the treatment of employee exhibiting symptoms compatible with a diagnosis of decompression illness; to be adjacent to an adequate emergency medical facility, which is equipped with demand-type oxygen inhalation equipment approved by the U.S. Bureau of Mines that is capable of being maintained at a temperature not to exceed 90 degree F., nor be less than 70 degree F; and to be provided with sources of air, free of oil and carbon monoxide, for normal and emergency use, which are capable of raising the air pressure in the lock from 0 to 75 psig. in 5 minutes.

Identifying Workers

Identification badges are to be furnished to all employees, indicating that the wearer is a compressed air worker. A permanent record is to be kept of all identification badges issued. The badge must give the employee's name, the address of the medical lock, the telephone number of the licensed physician for the compressed air project, and instructions in case there is an emergency of an unknown or doubtful cause, or illness, where the wearer may need to be rushed to the medical lock. The badge is to be worn at all times – off the job, as well as on the job.

Communications

Effective and reliable means of communication, such as bells, whistles, or telephones, are to be maintained, at all times, between all the following locations: the working chamber face, the working chamber side of the man lock near the door, the interior of the man lock, the lock attendant's station, the compressor plant, the first aid station, the emergency lock (if one is required), and the special decompression chamber, if one is required.

Signs and Records

The time of decompression is to posted in each manlock, using the format in Table 22. This form is to be posted in the manlock at all times.

Any code of signals used is to be conspicuously posted near workplace entrances, and other locations, as may be necessary, to bring them to the attention of all employees concerned. For each 8-hour shift, a record of the employees, who are employed under air pressure, is to be kept by an employee who remains outside the lock, near the entrance. This record must show the period each employee spends in the air chamber, and the time taken from decompression. A copy is to be submitted to the appointed physician after each shift.

Table 22

Time of Decompression Form

Time of Decompression for This Lock
___ pounds to ___ pounds in ___ minutes.
___ pounds to ___ pounds in ___ minutes.
(Signed by) _____(Superintendent)

Compression

Every employee going under air pressure for the first time is to be instructed on how to avoid excessive discomfort. During the compression of employees, the pressure is not to be increased to more than 3 psig. within the first minute. The pressure is to be held at 3 psig., and again at 7 psig. sufficiently long enough to determine if any employees are experiencing discomfort. After the first minute, the pressure is to be raised uniformly, and at a rate not to exceed 10 psi per minute. If any employee complains of discomfort, the pressure is to be held to determine if the symptoms are relieved. If after 5 minutes the discomfort does not disappear, the lock attendant is to gradually reduce the pressure until the employee signals that the discomfort has ceased. If the worker does not indicate that the discomfort has disappeared, the lock attendant is to reduce the pressure to atmospheric, and the employee is to be released from the lock. No employee is to be subjected to pressure exceeding 50 pounds per square inch, except in emergency.

Decompression

Decompression to normal condition is to be in accordance with the Standard Decompression Tables. In the event it is necessary for an employee to be in compressed air more than once in a 24-hour period, the appointed physician is responsible for the establishment of methods and procedures of decompression applicable to repetitive exposures. If decanting is necessary, the appointed physician must establish procedures before any employee is permitted to be decompressed by decanting methods. The period of time that an employee spends at atmospheric pressure, when the employee is between the decompression (following the shift) and recompression, is not to exceed 5 minutes.

Manlocks

Except in emergency, no employees employed in compressed air, are to be permitted to pass from the working chamber to atmospheric pressure until after decompression. The lock attendant in charge of a manlock is to be under the direct supervision of the appointed physician. He/she is to be stationed at the lock controls, on the free air side, during the period of compression and decompression, and must remain at the lock control station whenever there are individuals in the working chamber, or in the manlock. Except where air pressure in the working chamber is below 12 psig., each manlock is to be equipped with automatic controls which, through taped programs, cams, or similar apparatus, automatically regulates decompressions. It is also to be equipped with manual controls which permits the lock attendant to override the automatic mechanism in the event of an emergency. A manual control, which can be used in the event of an emergency, is to be placed inside the manlock. A clock, thermometer, and continuous recording pressure gauge, with a 4-hour graph, are to be installed outside of each manlock, and are to be changed prior to each shift's decompression. The chart is to be of sufficient size to register a legible record of variations in pressure within the manlock, and is to be visible to the lock attendant. A copy of each graph is to be submitted to the appointed physician after each shift. In addition, a pressure gauge clock and thermometer are also to be installed in each manlock. Additional fittings are to be provided so that test gauges may be attached whenever necessary. Except where air pressure is below 12 psig. and there is no danger of rapid flooding, all caissons having a working area greater than 150 square feet, and each bulkhead in tunnels of 14 feet or more in diameter, or equivalent area, must have at least two locks in perfect working condition, one of which is used exclusively as a manlock, the other, as a materials lock.

Where only a combination man-and-materials lock is required, this single lock is to be of sufficient capacity to hold the employees of two successive shifts.

Emergency locks are to be large enough to hold an entire heading shift and a limit maintained of 12 psig. There is to be a chamber available for oxygen decompression therapy to 28 psig. The manlock is to be large enough so that those using it are not compelled to be in a cramped position; it must not have less than 5 feet clear head room at the center; and it must have a minimum of 30 cubic feet of air space per occupant.

Locks on caissons are to be located so that the bottom door is not less than three feet above the water level surrounding the caisson on the outside. (The water level, where it is affected by tides, is construed to mean high tide.) In addition to the pressure gauge in the locks, an accurate pressure gauge is to be maintained on the outer and inner side of each bulkhead. These gauges are to be accessible at all times and are to be kept in accurate working order. Manlocks must have an observation port at least four inches in diameter, and they must be located in such a position that all occupants of the manlock may be observed from the working chamber, and from the free air side of the lock. Adequate ventilation in the lock is to be provided. Manlocks are to be maintained at a minimum temperature of 70 degree F. When locks are not in use and employees are in the working chamber, lock doors are to be kept open to the working chamber, where practicable. Provision is to be made to allow for rescue parties to enter the tunnel, if the working force is disabled.

Special Decompression Chamber

A special decompression chamber, of sufficient size to accommodate the entire force of employees being decompressed at the end of a shift, is to be provided whenever the regularly established working period requires a total time of decompression exceeding 75 minutes. The headroom in the special decompression chamber is not to be less than a minimum of 7 feet, and the cubical content must provide at least 50 cubic feet of airspace for each employee. For each occupant, there is to be provided 4 square feet of free walking area, and 3 square feet of seating space, exclusive of the area required for lavatory and toilet facilities. The rated capacity is to be based on the stated minimum space per employee, and is to be posted at the chamber entrance. The posted capacity is not to be exceeded, except in case of emergency. Each special decompression chamber is to be equipped with: a clock, or clocks, suitably placed so that the attendant and the chamber occupants can readily ascertain the time; pressure gauges which will indicate to the attendants, and to the chamber occupants, the pressure in the chamber; valves which will enable the attendant to control the supply and discharge of compressed air into and from the chamber. Valves and pipes, which are in connection with the air supply and exhaust, are to be arranged so that the chamber pressure can be controlled from within and without; an effective means of oral intercommunication between the attendant, occupants of the chamber, and the air compressor plant is to be supplied; and an observation port, at the entrance, is to be used to permit observation of the chamber occupants.

Seating facilities, in special decompression chambers, are to be so arranged as to permit a normal sitting posture without cramping. Seating space, not less than 18 inches by 24 inches wide, is to be provided, per occupant. Adequate toilet and washing facilities, in a screened or enclosed recess, are to be provided. Toilet bowls must have built-in protectors on the rim, so that an air space is created when the seat lid is closed. Fresh pure drinking water is to be available. This may be accomplished by either piping water into the special decompression chamber and providing drinking fountains, by providing individual canteens, or by some other sanitary means. Community drinking vessels are prohibited. No refuse or discarded material of any kind shall be permitted to accumulate, and the chamber is to be kept clean. Unless the special decompression chamber is serving as the manlock to atmospheric pressure, the special decompression chamber is to be situated, where practicable, adjacent to the manlock on the atmospheric pressure side of the bulkhead. A passageway is to be provided, connecting the special chamber with the manlock in order to permit employees, in the process of decompression, to move from the manlock to the special chamber, without a reduction in the ambient pressure from that designated for the next stage of decompression. The passageway is to be so arranged as to not interfere with the normal operation of the manlock, nor with the release of the occupants, of the special chamber, to atmospheric pressure upon the completion of the decompression procedure.

Compressor Plant and Air Supply

The compressor plant and air supply is to be manned, at all times, by a thoroughly experienced, competent, and reliable person; the competent person is to be at the air control valves as a gauge tender who regulates the pressure in the working areas. During tunneling operations, a gauge tender may regulate the pressure in two headings, but only if the gauges and controls are all in one location. In caisson work, there is to be a gauge tender for each caisson.

The low air compressor plant shall be of sufficient capacity to not only permit the work to be done safely, but also must provide a margin to meet emergencies and repairs. Low air compressor units must have at least two independent and separate sources of power supply, and each is to be capable of operating the entire low air plant, and its accessory systems. The capacity, arrangement, and number of compressors are to be sufficient enough to maintain the necessary pressure, without overloading the equipment, and they must also be sufficient enough to assure the maintenance of such pressure in the working chamber during periods of breakdown, repair, or an emergency.

Switching from one independent source of power supply to the other is to be done periodically, to ensure the workability of the apparatus in an emergency. Duplicate low-pressure air feedlines and regulating valves are to be provided between the source of air supply, and a point beyond the locks, with one of the lines extending to within 100 feet of the working face. All high- and low-pressure air supply lines are to be equipped with check valves. Low-pressure air is to be regulated automatically. In addition, manually operated valves are to be provided for emergency conditions. The air intakes for all air compressors are to be located at a place where fumes, exhaust, gases, and other air contaminants will be at a minimum. Gauges, indicating the pressure in the working chamber, are to be installed in the compressor building, the lock attendant's station, and at the employer's field office.

Compressed Air Ventilation and Air Quality

Exhaust valves and exhaust pipes are to be provided and operated so that the working chamber is well ventilated, and there are no pockets of dead air. Outlets may be required at intermediate points along the main low-pressure air supply line, to the heading, to eliminate such pockets of dead air. Ventilating air is not to be less than 30 cubic feet per minute. The air in the workplace is to be analyzed by the employer, not less than once each shift, and records of such tests are to be kept on file at the place where the work is in progress. The test results are to be within the threshold limit values for hazardous gases, and within 10 percent of the lower explosive limit of flammable gases. If these limits are not met, immediate action to correct the situation must be taken by the employer. The temperature of all working chambers which are subjected to air pressure shall, by means of after-coolers or other suitable devices, be maintained at a temperature not to exceed 85 degree F.

Forced ventilation is to be provided during decompression. During the entire decompression period, forced ventilation, through chemical or mechanical air purifying devices that will ensure a source of fresh air, is to be provided. Whenever heat-producing machines (moles, shields) are used in compressed air tunnel operations, a positive means of removing the heat build-up at the heading shall be provided.

All lighting in compressed-air chambers is to be by electricity, exclusively, and two independent electric-lighting systems, with independent sources of supply, are to be used. The emergency source is to be arranged to become automatically operative in the event of failure of the regularly used source. The minimum intensity of the light on any walkway, ladder, stairway, or working level is not to be less than 10 foot-candles, and in all workplaces the lighting is, at all times, to be such as to enable employees to see clearly. All electrical equipment and wiring for light and power circuits must comply with the requirements of construction electrical safety for use in damp, hazardous, high temperature, and compressed air environments. External parts of lighting fixtures and all other electrical equipment, when within 8 feet of the floor, are to be constructed of noncombustible, non-absorptive, insulating materials, except that metal may be used if it is effectively grounded. Portable lamps are to be equipped with noncombustible, non-absorptive, insulating sockets, approved handles, basket guards, and approved cords. The use of worn or defective portable and pendant conductors is prohibited.

Sanitation

Sanitary, heated, lighted, and ventilated dressing rooms and drying rooms are to be provided for all employees engaged in compressed air work. Such rooms must contain suitable benches and lockers. Bathing accommodations (showers at the ratio of one to 10 employees per shift), are to be equipped with running hot and cold water, and suitable and adequate toilet accommodations are to be provided. One toilet for each 15 employees, or fractional part thereof, is to be provided. When the toilet bowl is shut by a cover, there should be an air space so that the bowl or bucket does not implode when pressure is increased. All parts of caissons, and other working compartments, are to be kept in a sanitary condition.

Fire Prevention

Firefighting equipment is to be available at all times, and it shall be maintained in working condition. While welding or flame-cutting is being done in compressed air, a firewatch, with a fire hose or approved extinguisher, must stand by until such operation is completed. Shafts and caissons containing flammable material of any kind, either above or below ground, are to be provided with a waterline. A fire hose is to be connected so that all points of the shaft or caisson are within reach of the hose stream, and the fire hose must be at least 1 1/2 inches in nominal diameter. The water pressure must be, at all times, adequate for efficient operation of the type of nozzle used, and the water supply must be sufficient enough to ensure an uninterrupted flow. The fire hose, when not in use, is to be located in an area where it is guarded from damage.

The power house, compressor house, and all buildings which house ventilating equipment, are to be provided with at least one hose connection in the water line, with a fire hose connected thereto. A fire hose is to be maintained within reach of structures of wood over or near shafts.

Tunnels are to be provided with a 2-inch minimum diameter water line which extends into the working chamber, and it must be within 100 feet of the working face. Such lines must have hose outlets with 100 feet of fire hose attached, and one must be maintained at each of the following locations: at the working face, inside the bulkhead of the working chamber, and one immediately outside the bulkhead. In addition, hose outlets are to be provided at 200-foot intervals throughout the length of the tunnel, and 100 feet of fire hose is to be attached to the outlet nearest any location where flammable material is being kept or stored, or where any flame is being used.

In addition to the required fire hose protection, every floor of every building which is not under compressed air work, but is used in connection with the compressed air work, is to be provided with at least one approved fire extinguisher of the proper type for the hazard involved. At least two approved fire extinguishers are to be provided in the working chamber, with one at the working face and one immediately inside the bulkhead (pressure side). Extinguishers which are used in the working chamber, and use water as the primary extinguishing agent, must not use any extinguishing agent which could be harmful to the employees in the working chamber. The fire extinguisher is to be protected from damage.

Highly combustible materials are not to be used or stored in the working chamber. Wood, paper, and similar combustible material, are not to be used in the working chamber in quantities which could cause a fire hazard. The compressor building must be constructed of noncombustible material.

Manlocks are to be equipped with a manual type fire extinguisher system that can be activated inside the manlock, and also outside by the lock attendant. In addition, a fire hose and portable fire extinguisher are to be provided inside and outside the manlock. The portable fire extinguisher is to be the dry chemical type.

Equipment, fixtures, and furniture in manlocks and special decompression chambers are to be constructed of noncombustible materials. Bedding, etc. is to be chemically treated so as to be fire resistant. Head frames are to be constructed of structural steel, or open frame-work fireproofed timber. Head houses and other temporary surface buildings or structures that are within 100 feet of the shaft, caisson, or tunnel opening are to be built of fire-resistant materials.

No oil, gasoline, or other combustible material is to be stored within 100 feet of any shaft, caisson, or tunnel opening, except oils may be stored in suitable tanks, in isolated fireproof buildings, provided the buildings are not less than 50 feet from any shaft, caisson, or tunnel opening, or any building directly connected to underground operations. Positive means are to be taken to prevent leaking flammable liquids from flowing into the areas previously mentioned. All explosives used in connection with compressed air work are to be selected, stored, and transported according to the standard rules for underground use and storage.

Bulkheads and Safety Screens

Intermediate bulkheads, with locks or intermediate safety screens, or both, are to be required where there is the danger of rapid flooding. In tunnels 16 feet or more in diameter, hanging walkways are to be provided from the face to the manlock, and as high in the tunnel as practicable, with at least 6 feet of head room. Walkways are to be constructed of noncombustible material. Standard railings are to be securely installed, on open side, throughout the length of all walkways. Where walkways are ramped under safety screens, the walkway surface is to be skidproofed by cleats, or by equivalent means. Bulkheads used to contain compressed air, shall be tested, where practicable, to prove their ability to resist the highest air pressure which may be expected to be used.

VERMIN CONTROL (1926.51)

In workplaces where there are rodents, insects, and other vermin, a continuous extermination program is to be used.

WASHING FACILITIES (1926.51)

The employer must provide adequate washing facilities for employees engaged in the application of paints, coating, herbicides, or insecticides, or in other operations where contaminants may be harmful to the employees. Such facilities are to be in near proximity to the worksite, and are to be equipped to enable employees to remove such substances. Washing facilities shall be maintained in a sanitary condition.

Lavatories are to be made available in all places of employment. These requirements do not apply to mobile crews, or to normally unattended work locations, if employees working at these locations have transportation readily available to them to nearby washing facilities, and these facilities meet the other requirements of this paragraph. Each lavatory is to be provided with hot and cold running water, or tepid running water. Hand soap, or similar cleansing agents, are to be provided; as well as individual hand towels, sections of cloth or paper towels, clean, individual sections of continuous cloth toweling, or warm air blowers, convenient to the lavatories, must be provided.

Whenever showers are required by a particular standard, one shower is to be provided for each 10 employees of each sex, or numerical fraction thereof, who are required to shower during the same shift.

Body soap, or other appropriate cleansing agents, convenient to the showers, must be provided. Showers are to be provided with hot and cold water which feeds a common discharge line. Employees who use showers are to be provided with individual clean towels.

WELDING (1926.350)

Only experienced persons are allowed to do electrical or acetylene welding or cutting. No welding or burning is to be done in hazardous areas without a permit. Warning signs, signaling overhead welding and cutting, must be posted. No welding or cutting should be done on barrels or tanks. Special precautions need to be taken while welding or cutting in a confined space. When welding, workers must wear the proper eye and face protection. Welders may be subjected to the inhalation of toxic fumes, which can cause illness; and they may also be subject to safety hazards, such as fire, which could result in a fatality, serious injury, and/or property damage; therefore, special precautions need to be followed when welding operations are in progress. See Table 23 for a summary of welding requirements.

Fire Prevention (1926.352)

When practical, objects to be welded, cut, or heated are to be moved to a designated safe location or, if the objects to be welded, cut, or heated cannot be readily moved, all movable fire hazards in the vicinity are to be taken to a safe place, or otherwise protected. If the object to be welded, cut, or heated cannot be moved, and if all the fire hazards cannot be removed, positive means must be taken to confine the heat, sparks, and slag, and to protect the immovable fire hazards from them. No welding, cutting, or heating is to be done where the application of flammable paints, or the presence of other flammable compounds, or heavy dust concentrations create a hazard. Suitable fire extinguishing equipment is to be immediately available in the work area, and is to be maintained in a state of readiness for instant use.

When the welding, cutting, or heating operation is such that normal fire prevention precautions are not sufficient, additional personnel must be assigned to guard against fire while the actual welding, cutting, or heating operation is being performed, and the additional personnel must remain for a sufficient period of time, after the completion of the work, to ensure that no possibility of fire exists. Such personnel are to be instructed as to the specific anticipated fire hazards, and how the firefighting equipment provided is to be used.

When welding, cutting, or heating is performed on walls, floors, and ceilings, and since direct penetration of sparks or heat transfer may introduce a fire hazard to an adjacent area, the same precautions are to be taken on the opposite side, as are taken on the side on which the welding is being performed.

For the elimination of possible fire in enclosed spaces, that are a result of gas escaping through leaking or improperly closed torch valves, the gas supply to the torch is to be positively shut off at a point which is outside the enclosed space; this should be done whenever the torch is not to be used, or whenever the torch is left unattended for a substantial period of time, such as during the lunch period. Overnight, and at the change of shifts, the torch and hose are to be removed from the confined space. Open-end fuel gas and oxygen hoses are to be immediately removed from enclosed spaces when they are disconnected from the torch, or other gas-consuming device.

Except when the contents are being removed or transferred, drums, pails, and other containers, which contain or have contained flammable liquids, are to be kept closed. Empty containers are to be removed to a safe area apart from hot work operations or open flames.

Before welding, cutting, or heating is undertaken on drum containers, or hollow structures which have contained toxic or flammable substances, they are to either be filled with water, or thoroughly cleaned of such substances, and ventilated and tested. For welding, cutting, and heating on steel pipelines containing natural gas, the pertinent portions of regulations issued by the Department of Transportation, Office of Pipeline Safety, 49 CFR Part 192, Minimum Federal Safety Standards for Gas Pipelines, apply. Before heat is applied to a drum, container, or hollow structure, a vent or opening is to be provided for the release of any built-up pressure during the application of heat.

Ventilation and Protection in Welding, Cutting, and Heating (1926.353)

Mechanical ventilation must meet the following requirements:

1. Mechanical ventilation must consist of either general mechanical ventilation systems or local exhaust systems.

Table 23

Summary of Welding Requirements

General Requirements

1. Only qualified welders are to be authorized to do any welding, heating, or cutting.

2. Inspect your work area for fire hazards and proper ventilation before welding or cutting.

3. Avoid welding or cutting sparks, and hot slag. Be alert to hot surfaces, and avoid touching metal surfaces until they have cooled.

4. Place compressed gas cylinders in an upright position, and secure them in place to prevent dropping or falling. Handle them with extreme care, and do not store them near any sources of heat.

5. Remove any combustibles when welding or cutting must be done. If removal is not feasible, cover combustibles with a noncombustible material. When welding near any combustible material, another employee must be posted to serve as a fire watch. Make sure this person has a fire extinguisher available, and keep him/her in the area after welding/cutting is completed, and until all danger of fire is past.

6. When working in the vicinity of welding operations, wear approved goggles and avoid looking directly at the flash, as serious flash burns could result.

7. When opening valves on tanks that have regulators installed, be sure the pressure adjustment screw is all the way out, and do not stand in front of the regulator. An internal failure could rupture the regulator and cause the adjustment screw to become a missile.

Gas Welding and Cutting

1. When transporting, moving, and storing compressed gas cylinders, always ensure that the valve protection cap is in place and secured.

2. Secure cylinders on a cradle, slingboard, or pallet when hoisting. Never hoist or transport by means of magnet or choker slings.

3. Move cylinders by tilting and rolling them on their bottom edges. Do not allow cylinders to be dropped, struck, or come into contact with other cylinders, violently.

4. Secure cylinders in an upright (vertical) position, when transporting by powered vehicles.

5. Do not hoist cylinders by lifting on the valve protection caps.

6. Do not use bars under valves, or valve protection caps, to pry cylinders loose when frozen. Use warm, not boiling, water to thaw cylinders loose.

7. Remove regulators and secure valve protection caps prior to moving cylinders, unless cylinders are firmly secured on a special carrier intended for transport.

8. Close the cylinder valve when work is finished, when cylinders are empty, or when cylinders are moved at any time.

Table 23

Summary of Welding Requirements (*Continued*)

9. Secure compressed gas cylinders in an upright position (vertical), except when cylinders are actually being hoisted or carried.

Arc Welding and Cutting

1. Use only manual electrode holders which are specifically designed for arc welding and cutting.
2. All current-carrying parts, passing through the portion of the holder, must be fully insulated against the maximum voltage encountered to ground.
3. All arc welding and cutting cables must be completely insulated, flexible type, and capable of handling the maximum current requirements of the work in progress.
4. Report any defective equipment to your supervisor, immediately, and refrain from using such equipment.
5. Shield all arc welding and cutting operations, whenever feasible, by noncombustible or flameproof screens, to protect employees and other persons working in the vicinity from the direct rays of the arc.

Fire Prevention

1. Locate the nearest fire extinguisher in your work area in case of future need for an emergency. Fire extinguishing equipment must be immediately available in the work area.
2. Never use matches or cigarette lighters. Use only friction lighters to light torches.
3. Never strike an arc on gas cylinders.
4. Move objects to be welded, cut, or heated to a designated safe location. If the objects cannot readily moved, then all movable fire hazards, in the vicinity, must be taken to a safe place, or otherwise protected.
5. Do not weld, cut, or heat where the application of flammable paints, or the presence of other flammable compounds, or heavy dust concentrations creates a hazard.
6. Additional employees must be assigned to guard against fire, while the actual welding, cutting, or heating is being performed, when the operation is such that normal fire prevention precautions are not sufficient.
7. Prior to applying heat to a drum, container, or hollow structure, provide a vent or opening to release any built-up pressure during the application of heat.
8. Never cut, weld, or heat on drums, tanks, or containers that have contained flammable liquids, until they have been cleaned.

2. General mechanical ventilation is to be of sufficient capacity and so arranged as to produce the number of air changes necessary to maintain welding fumes and smoke within safe limits.

3. Local exhaust ventilation must consist of freely movable hoods, intended to be placed by the welder or burner, as close as practicable to the work. This system must be of sufficient capacity and so arranged as to remove fumes and smoke at the source, and keep the concentration of them, in the breathing zone, within safe limits.

4. Contaminated air, exhausted from a working space, is to be discharged into the open air, or otherwise clear of the source of intake air.

5. All air replacing shall be clean and respirable.

6. Oxygen is not to be used for ventilation purposes, comfort cooling, blowing dust from clothing, or for cleaning the work area.

General mechanical, or local exhaust ventilation is to be provided whenever welding, cutting, or heating is performed in a confined space. When sufficient ventilation cannot be obtained without blocking the means of access, employees in the confined space are to be protected by air line respirators. An employee, on the outside of such a confined space, is to be assigned to maintain communication with those working within it, and the employee is to aid them in an emergency. When a welder must enter a confined space through a manhole or other small opening, means are to be provided for quickly removing him in case of emergency. When safety harnesses and lifelines are used for this purpose, they are to be attached to the welder's body so that his body cannot be jammed in a small exit opening. An attendant, with a preplanned rescue procedure, is to be stationed outside to observe the welder, at all times, and is to be capable of putting rescue operations into effect.

Welding, cutting, or heating, in any enclosed space which involves metals that have zinc-bearing bases, filler metals, metals coated with zinc-bearing materials, lead base metals, cadmium-bearing filler materials, chromium-bearing metals, or metals coated with chromium-bearing materials, is to be performed with either general mechanical or local exhaust ventilation.

Any welding, cutting, or heating, in any enclosed space involving the following metals: metals containing lead, other than as an impurity, or metals coated with lead-bearing materials, cadmium-bearing or cadmium-coated base metals, metals coated with mercury-bearing metals, beryllium-containing base or filler metals shall be done with ventilation and air-line respirators. Because of its high toxicity, work involving beryllium is done with both local exhaust ventilation and air line respirators. All other welding, cutting, or heating of these metals is to be performed with local exhaust ventilation, or the employees shall be protected by air line respirators.

Employees performing such operations in the open air are to be protected by filter-type respirators, except that employees performing such operations on beryllium-containing base, or filler metals are to be protected by air line respirators. Other employees, who are exposed to the same atmosphere as the welders or burners, are to be protected in the same manner as the welders or burners.

Since the inert-gas metal-arc welding process involves the production of ultra-violet radiation of intensities of 5 to 30 times that produced during shielded metal-arc welding, the decomposition of chlorinated solvents by ultraviolet rays, and the liberation of toxic fumes and gases, the employees are not to be permitted to engage in, or be exposed to the process until the following special precautions have been taken. The use of chlorinated solvents are to be kept at least 200 feet, unless shielded, from the exposed arc, and the surfaces prepared with chlorinated solvents, are to be thoroughly dry before welding is permitted on such surfaces. Employees in the area, not protected from the arc by screening, shall be protected by filter lenses. When two or more welders are exposed to each other's arc, filter lens goggles of a suitable type, are to be worn under welding helmets. Hand shields, to protect the welder against flashes and radiant energy, are to be used when either the helmet is lifted, or the shield is removed. Welders, and other employees who are exposed to radiation, are to be suitably protected so that the skin is completely covered, to prevent burns and other damage by ultraviolet rays. Welding helmets, and hand shields, are to be free of leaks and openings, and free of highly reflective surfaces.

When inert-gas metal-arc welding is being performed on stainless steel, workers are to be protected against dangerous concentrations of nitrogen dioxide.

Welding, cutting, and heating, involving normal conditions or materials, may be done without mechanical ventilation or respiratory protective equipment, but where, because of unusual physical or atmospheric conditions, an unsafe accumulation of contaminants exists, suitable mechanical ventilation or respiratory protective equipment shall be provided. Employees performing any type of welding, cutting, or heating are to be protected by suitable eye protective equipment.

Welding, Cutting, and Heating of Preservative Coatings (1926.354)

Before welding, cutting, or heating commences on any surface covered by a preservative coating and the flammability is not known, a test is to be made by a competent person to determine its flammability. Preservative coatings are to be considered highly flammable when scrapings burn with extreme rapidity. Precautions are to be taken to prevent ignition of highly flammable, hardened preservative coatings. When coatings are determined to be highly flammable, they are to be stripped from the area to be heated to prevent ignition. In enclosed spaces, all surfaces covered with toxic preservatives are to be stripped of all toxic coatings for a distance of at least 4 inches from the area of heat application, or the employees are to be protected by air line respirators. When the previous task is conducted in the open air, employees are to be protected by a respirator. To ensure that the temperature of the unstripped metal will not be appreciably raised, the preservative coatings are to be removed at a sufficient distance from the area to be heated. Artificial cooling of the metal surrounding the heating area may be used to limit the size of the area required to be cleaned.

WOODWORKING TOOLS (1926.304)

All woodworking tools and machinery must meet other applicable requirements of the American National Standards Institute, 01.1-1961, Safety Code for Woodworking Machinery. All fixed power-driven woodworking tools are to be provided with a disconnect switch that can either be locked or tagged in the off position. The operating speed is to be etched, or otherwise permanently marked on all circular saws over 20 inches in diameter, or operating at over 10,000 peripheral feet per minute. Any saw so marked is not to be operated at a speed other than that marked on the blade. When a marked saw is retensioned for a different speed, the marking is to be corrected to show the new speed.

Automatic feeding devices are to be installed on machines, whenever the nature of the work will permit. Feeder attachments must have the feed rolls, or other moving parts covered or guarded so as to protect the operator from hazardous points.

All portable, power-driven circular saws are to be equipped with guards above and below the base plate or shoe. The upper guard must cover the saw to the depth of the teeth, except for the minimum arc required to permit the base to be tilted for bevel cuts. The lower guard must cover the saw to the depth of the teeth, except for the minimum arc required to allow proper retraction and contact with the work. When the tool is withdrawn from the work, the lower guard must be automatically and instantly returned to the covering position.

On radial saws, the upper hood must completely enclose the upper portion of the blade, down to a point that will include the end of the saw arbor. The upper hood is to be constructed in such a manner, and of such material, that it will protect the operator from flying splinters, broken saw teeth, etc. and will defect sawdust away from the operator. The sides of the lower exposed portion of the blade are to be guarded to the full diameter of the blade, by a device that will automatically adjust itself to the thickness of the stock, and it must remain in contact with the stock being cut, to give the maximum protection possible for the operation being performed.

Each circular crosscut table saw is to be guarded by a hood which meets all the requirements for the hoods for circular ripsaws. Each circular hand-fed ripsaw is to be guarded by a hood which completely encloses the portion of the saw above the table, and the portion of the saw above the material being cut. The hood and mounting are to be arranged so that the hood will automatically adjust itself to the thickness of, and remain in contact with, the material being cut, but it must not offer any considerable resistance to insertion of the material to the saw, or to passage of the material being sawed. The hood is to be made of adequate strength to resist blows and strains incidental to reasonable operation, adjusting, and handling, and is to be so designed as to protect the operator from flying splinters and broken saw teeth. It is to be made of material that is soft enough so that it will be unlikely to cause tooth breakage. The hood is to be so mounted as to ensure that its operation will be positive, reliable, and in true alignment with the saw; and the mounting shall be adequate in strength to resist any reasonable side thrust, or other force tending to throw it out of line.

Workers should wear eye protection, hearing protection, and other personal protective equipment, as is appropriate to protect them from injury.

WORKING OVER WATER (1926.106)

Many of the deaths which occur while working around or over water are due to the failure to wear a U.S. Coast Guard approved life jacket. The life jacket is often found near the scene of the accident, or a life jacket was not supplied by the employer. Even when the victim is a swimmer and falls into the water, the victim may be injured by the fall, may strike another object, may find the current too strong, or may find the water too cold, making survival difficult.

Precautions, prior to an occurrence, are the easiest prevention. The simple solution to this problem is to require life jackets to always be worn when working over or around water. It is also important to provide a life boat for rescue purposes, as well as ring buoys with 90 feet of line, spaced 200 feet apart.